THE PROTEUS EFFECT

Other titles by Ann B. Parson

Menopause, coauthored with Isaac Schiff

Decoding Darkness: The Search for the Genetic Causes of Alzheimer's Disease, coauthored with Rudolph E. Tanzi

THE PROTEUS EFFECT

STEM CELLS AND THEIR PROMISE FOR MEDICINE

Ann B. Parson

Joseph Henry Press
Washington, D.C.

Joseph Henry Press • 500 Fifth Street, NW • Washington, DC 20001

The Joseph Henry Press, an imprint of the National Academies Press, was created with the goal of making books on science, technology, and health more widely available to professionals and the public. Joseph Henry was one of the founders of the National Academy of Sciences and a leader in early American science.

Any opinions, findings, conclusions, or recommendations expressed in this volume are those of the author and do not necessarily reflect the views of the National Academy of Sciences or its affiliated institutions.

Library of Congress Cataloging-in-Publication Data

Parson, Ann B.
  The Proteus effect : stem cells and their promise for medicine / by Ann B. Parson.
      p. ; cm.
  Includes bibliographical references and index.
  ISBN 0-309-08988-3 (cloth with jacket) ISBN 0-309-53329-5 (PDF)
  1. Stem cells—Research—History. 2. Stem cells—Popular works.
  [DNLM: 1. Research—history. 2. Stem Cells. 3. Stem Cell
Transplantation. QH 581.2 P266p 2004] I. Title.
  QH588.S83P37 2004
  616'.02774—dc22
                                       2004013757

Cover image: © Victor Habbick Visions/Photo Researchers, Inc.

Printed in the United States of America.

First Printing September, 2004
Second Printing November, 2004

For my mother,
who showed me color

For Streaker,
who basks in the natural

Contents

Proteus—a Sea God and Ocean Shepard for Poseidon's seals—was able to change shape when the occasion warranted.

Some have the gift to change and change again in many forms, like Proteus, creature of the encircling seas, who sometimes seemed a lad, sometimes a lion, sometimes a snake men feared to touch, sometimes a charging boar, or else a sharp-horned bull; often he was a stone, often a tree, or feigning flowing water seemed a river or water's opposite a flame of fire.

—Metamorphoses 8.731

To Proteus. Proteus I call, whom fate decrees to keep the keys which lock the chambers of the deep; first-born, by whose illustrious power alone all nature's principles were clearly shown. Pure sacred matter to transmute is thine, and decorate with forms all-various and divine. All-honoured, prudent, whose sagacious mind knows all that was and is of every kind, with all that shall be in succeeding time, so vast thy wisdom, wondrous and sublime: for all things Nature first to thee consigned, and in thy essence omniform confined. O father, to the mystics' rites attend, and grant, a blessed life a prosperous end.

—*Orphic Hymn 25 to Proteus*

Introduction

The call went out from a Subaru wagon as it rapidly approached Boston on a crowded Massachusetts turnpike. It was an April morning in 2001. At the downtown offices of the cord blood banking company Viacord, which were located two floors above the Hard Rock Cafe in Boston's russet Back Bay neighborhood, one of the twenty-five phones in the calling center lit up and was promptly answered. Viacord's operators were on tap twenty-four hours a day for their clients, far outdoing the bar below, where the only light that had remained on all night was the blood-red neon sign in the window: "No Drugs or Nuclear Weapons Allowed Inside."

"My wife's in labor; we're on our way to the Brigham," said the caller, referring to the Brigham and Women's Hospital. "Yes, we remembered the kit. It's in the trunk."

The caller was George Daley. No stranger to hospitals, he himself was an M.D., a hematologist-oncologist who had served as chief resident in medicine at Massachusetts General Hospital some years earlier. Approaching midpoint in his career, he was in an enviable position, spending a portion of his time attending to the needs of cancer patients at Mass General and the rest of his time developing strategies against a rare form of leukemia in his laboratory at the prestigious Whitehead Institute across the Charles River. As for "the kit," every Viacord customer received one. They in turn had to remember to give it to their obstetrician or midwife right before their newborn entered the world. Resembling a square shoe box, it contained several items that facilitated salvaging blood from an infant's umbilical cord and attached placenta for the sake of storing this natural resource long-term. Included were a standard blood-collection bag, an umbilical cord clamp, a two-inch needle for drawing the blood from the cord, and an anticoagulant to ensure the bagged blood wouldn't clot, which would have ruined the usefulness of certain of its cells. The whole point of bagging the blood was to save these special cells.

Ideally, most of little Nicholas Daley's cord blood—about three ounces' worth—would end up frozen in a blood-storage facility several hundred miles away that Viacord leased from the University of Cincinnati. The sample might very well sit there unused for the rest of Nicholas's life. As far as his parents were concerned, that would be their wish. Tucked away, the blood would be medical insurance at its most basic, and were it ever needed, it would be because Nicholas, a sibling, or some other relative had encountered a serious, life-threatening illness, and the hope at that point would be that the special cells in Nicholas's cord blood, which floated in rare numbers among other types of blood cells, would come to the rescue.

The cells held in such high regard were stem cells—in this case, stem cells specific to a baby's cord blood and therefore a shade more differentiated, or specialized, than the stem cells present in early development, in the early embryo. In either case, what generally sets a stem cell apart from other cells is that it is of immature status (some versions more than others) and unspecialized (some versions more than others) and more fancy-free than a mature cell that has specialized.

When it divides, a stem cell can be of two minds; it can generate more stem cells just like itself—scientists refer to this as "self-renewing"—or it can produce specialized fare. This is said to be why scientists call it a stem cell—because of the different cells that can *stem* from it. While mature cells—the heart's myocytes or the liver's hepatocytes or epithelial cells in the lungs—are apt to stop dividing after several dozen divisions, the beauty of stem cells is that they can keep dividing indefinitely, renewing themselves and/or churning out differentiated progeny. As a result, they are terribly important tissue regenerators. Notes ophthalmologist Kenneth Kenyon, who later enters our story, "The essence of a stem cell is that it is a gift that keeps on giving." Without stem cells, and the specialized tissue they provide, says Kenyon, "we'd be short-lived, short-limbed, and short-sighted as well." We'd be nothing, actually, nor would any other multicellular living thing.

Of the several different versions of stem cells that Nicholas Daley would benefit from before and after birth, the one that begins it all is the fertilized egg, or zygote, which scientists consider the ultimate, or "mother," stem cell. The zygote is *totipotent*, meaning that it has the capacity to give rise to every type of cell that festoons every tissue associated with the embryo and eventual adult, including the umbilical and placental structures that support the growing embryo. Because a totipotent stem cell has total potential to become any cell, it has what is referred to as unbounded "stemness."

Even after the fertilized egg cleaves once, then twice, its multiplying cells continue to be totipotent, although in the space of a few more divisions this magic appears to be gone. Starting on day 4 postfertilization, however, just as it reaches the uterus but before it implants therein, the early human embryo acquires another category of stem cell. As it evolves into a hollow round ball—a blastocyst—a little heap of cells appears on its inner wall. This small clump, which represents the first glimpse of the embryo proper, is composed of stem cells that are considered *pluripotent* by most scientists. Whereas the totipotent (*totus* meaning entire) version of a cell can generate every cell type associated with the growing individual, the pluripotent (*plures* meaning several) version nearly can, the exception being the cells that make up the embryo's outer layer (the trophoblast) and the embryo-supporting placenta that forms from those cells.

The stem cells heaped inside the blastocyst, or early embryo, look not unlike a clump of seeds inside a round pod. But each of these cells is much tinier than a plant seed. On day 5 post-fertilization, the human blastocyst itself "is smaller than a period set in six-point type," describes William Lensch, a stem cell scientist in George Daley's lab, now located at Children's Hospital in Boston. For these embryonic stem cells to be seen with any clarity, they need to be magnified by as much as five hundred times. Yet as extraordinarily small as an embryo-residing pluripotent stem cell is, it is astonishingly powerful. Through an exhaustive process of division and differentiation, or specialization, the blastocyst's inner handful of stem cells will grow into a bigger and bigger bundle of cells that ultimately arrange themselves into organized tissue and an individual.

By the time an embryo becomes a fetus, organs are beginning to form, and yet a third category of stem cell is coming into existence: *adult stem cells.* (In the case of human development, biologists usually cite week 8 post-fertilization as the beginning of the fetal stage.) Many an organ will have its own resident stem cells. So, too, as we've seen, do the umbilical cord and placenta. Stem cells in organs are few and far between—in some cases as rare as one in every 15,000 cells. Often, however, they will bear the responsibility of regenerating an organ's entire population of specialized cells. As the fetus grows, one particularly striking example of organ-dwelling stem cells are those that migrate into the developing testis or ovary, for these precursors of sperm and eggs cells will perpetuate the species. The outcome is one of Nature's most miraculous stunts: A highly specialized egg from a female will join with a highly specialized sperm from a male to produce a zygote, an unspecialized but fabulously totipotent cell that will bring about a new individual. Then round the circle will go again.

His wife's pregnancy had made George Daley think hard about this amazing phenomenon. Speaking at a gathering at Whitehead Institute right around the time that his son Nicholas was born, Daley noted how remarkable it was that a sperm from his "aged, decrepit body"—leanly built and smartly tailored, forty-year-old Daley hardly appeared decrepit—could fuse with an egg from his wife to bring about "a cell that is completely rejuvenated." Each of us might age and cease to be, but as a species, "we're immortal," he was led to conclude. Daley's topic that day was the future of stem cells in medicine, and in addressing a roomful of journalists, he swung between exuberant futuristic descriptions of cell-based therapies and cautious add-ons about how long this all could take. He couldn't hide his fervent belief, however, that stem cells would increasingly emerge as the basis for new, more rational medicines.

Once little Nicholas entered the world, his muscles, his skin, his brain, his blood, his intestines, and certain other organs would retain rare populations of adult stem cells that would continue to replenish his tissues for the rest of his life. Differentiated cells such as the brain's hippocampal cells or bone's osteocytes carry out the body's innumerable special tasks. Yet trace any mature cell back to its origins, and you always arrive at a stem cell, which, by the way, is the strongest argument to be made for what medicine stands to gain. The fact that stem cells are so pivotal in forming and maintaining tissues is seen as proof of their usefulness in medicine, and the best thing about them is that no chemist has to spend a lifetime at the bench coming up with such an ideal salve, because Nature already has done so. Some organ stem cells produce only one type of specialized cell. Those in the testis, for example, yield only sperm cells. But most adult stem cells are *multipotent*, meaning they produce many cell types. For instance, the hematopoietic (or blood-forming) stem cell that hovers in a person's bone marrow constantly regenerates the blood and immune systems' eight or so specialized cells. (One of many unsolved mysteries is, how do stem cells know how many mature cells to make or replace?)

In Daley's eyes, the blood-forming stem cell, around which so much of his lab's cancer research revolved, was without a doubt "the granddaddy of all stem cells," at least from a research standpoint. Scientists had fastened on it ahead of other versions of stem cells and pushed it into the limelight as a model for all stem cells—what they are like and how they behave. To physicians, moreover, the hematopoietic stem cell in bone marrow was already a distinguished hero. It, and it alone, was the reason that a bone marrow transplant, usually in combination with radiation or chemotherapy, had the ability to effectively treat many blood-related disorders, among them malignancies like leukemia, immune disorders like lupus, and genetic disorders like severe combined immunodeficiency disease (SCID) and sickle cell anemia. It was easy to have faith in this cell's therapeutic virtues, when, by the new millennium, patients receiving a bone marrow transplant sometimes had a fifty percent or better chance for improvement, depending on the blood-borne disease.

Yet three years earlier in 1998, Daley, a prominent member of the still-small stem cell biology field, had been distrustful of the stem cell that instead floated in an infant's umbilical cord blood. Without a doubt, it was related to the stem cell found in a baby's or adult's bone marrow. But whether the cord type could reestablish the blood system as successfully as the marrow type was unclear. In mice, it was possible to restore a full complement of red blood cells, white blood cells, and platelets—virtually the entire blood and immune systems—with just a handful of bone marrow stem cells. Did the umbilical cord's stem cell have the same ability?

When the Daleys' first child was due, several things fell into place that led them to preserve Jack's cord blood, despite the unknowns. George Daley's interest in cord blood's immature cells was growing as the evidence built that these hematopoietic cells, just like those in bone marrow, could breathe life back into destitute blood and immune systems. By the turn of the millennium, over 2,000 patients had received cord blood transplants, many saved from fatal illnesses. Studies were even suggesting that the cord variety of stem cell, when given to an unrelated recipient, provoked less of a negative immune reaction than the marrow variety. Cord blood stem cells apparently were so immature, they weren't readily detected by the recipient's patrolling lymphocytes. Daley in the meantime had accepted an invitation to become the scientific advisor of T Breeders, a biotech company that was attempting to grow and expand stem cells, and, as a gift, the company's founder offered the Daleys the opportunity of having their firstborn's cord blood stored by Viacord, a cord blood salvaging company that was betting its existence on the stem cells in cord blood.

By 2001, with his second son's birth fast approaching, Daley's doubts about the value of preserving cord blood had vanished. "This time I was completely, solidly behind the whole effort," he recalls. The stem cells that course through the umbilical cord and placenta might not be identical to those in bone marrow, but for the purposes of transplant therapies they were close enough.

Others in the research community and beyond have now come to the same conclusion, as indicated by the Cord Blood Stem Cell Act of 2003. Introduced into Congress by a bipartisan group of U.S. senators, the bill proposes establishing a National Cord Blood Stem Cell Bank Network that would work to collect enough cord blood units from births all over the country to provide an immunologic match for ninety percent of people needing a cord blood stem cell transplant.

This isn't to say, Daley acknowledges, that cord blood transplantation doesn't continue to face sizable hurdles. For instance, although a large majority of cord blood recipients ultimately engraft—the transplanted cells start dividing and generating more cells—"a significant percent of these patients die during the therapy itself because they don't engraft soon enough," says Daley. This happens, it seems, because so few cells in cord blood are the stem variety. The solution might be to infuse larger populations of cord stem cells into recipients. In the meantime, Daley is just glad to know that both of his sons' cord blood cells are safely stored away in some dark, cool place where they will escape the unhealthy mutational changes that the outside environment inflicts on cells. This way, these regenerative cells should remain fresh as daisies—or "pristine," as Daley observes.

Nicholas Edmondson Daley was born at 4:13 on the afternoon of April 30, 2001, one of the four million births that take place every year in the United States. But unlike what happens over ninety-seven percent of the time, his cord blood did not wind up in the "red bag," the receptacle into which the rest of the umbilical cord and placenta, along with the bloody gauzes, tonsils, IV lines, and other potentially infectious waste collected from throughout the hospital would get thrown for special handling and disposal. A few minutes after his son's birth, George Daley himself took the long needle from Viacord's kit and inserted it into the large vein that spirals down the length of the umbilical cord, draining the blood into a collection bag. The cord blood industry's refrain must have been running through his mind: *What a deplorable waste if a baby's cord blood ends up in a bag in some hospital basement, when the special cells in its midst might save a life.* A phone operator at Viacord was soon notified of the birth and in turn sent a courier to pick up the blood sample. Today, suspended in liquid nitrogen at a cool −196°C, these cells wait out an indefinite future.

What are these regenerative cells all about? How much promise do they really hold for medicine? One way to measure the future is to first search back through the past and take stock of how humans have gradually awakened to these distinctive, often camouflaged, cells in our midst and slowly come to recognize their worth.

What follows, then, is a continuum of experimenters and experiments that starts as far back as 1740 and proceeds into the present. In gathering together these threads, I've largely looked at stem cells in terms of where they might lead our medical community. These cells are such stunning creations, however, that anyone stopping to peer at them can't help but admire them for qualities that go far beyond their uses as simply tools for human medicine. Stem cells are basic to the regeneration of every multicellular plant and animal, and as scientists discover more about them, these flexible worlds-unto-themselves should open our eyes to the presence of forces in Nature that are far greater than anything humans could imagine or invent. As neuroscientist Evan Snyder has aptly put it, "Even the dumbest stem cell is smarter than the smartest scientist."

Botanists have actually been probing stem cells in plants for much longer than zoologists have studied stem cells in animals. They refer to them as "meristem cells" as opposed to stem cells, which helps to distinguish them from cells in a plant's actual stem. (*Meristem* comes from the Greek word *meristos*, for "divided.") Similar to the animal kind, meristem cells exist in plant embryos and are present as well in the organs of an adult plant. Roots have clusters of meristem cells at their tips, and as these cells divide, roots inch out through the soil, notes Susan Singer, a biologist at Carleton College. Above ground, self-perpetuating meristems at the tips of shoots attend to a plant's dazzling ability to branch and leaf. Flowers meanwhile bloom due to floral meristems, while trees add girth because of meristem cells in their cambria. Grasses such as corn have another type of meristem in their stalks that contributes to their elongation, says Singer. "If you go out into a field on a summer's night when the corn is still quite short and hear a popping sound, it's because of the really rapid growth of those cells."

On the whole, it's thanks to their stem cells that plants are such outstanding specimens of regeneration. Because it is rooted, a plant depends mightily on its regenerative growth for flexibility, notes Singer. For instance, its regrowth can be crucial for its reproductive life. "A plant wants to flower when its pollinators are around," says Singer, and it will "produce floral meristems at just the right time," with environmental cues like light and temperature indicating the right season. Regrowth also lends to a plant's survival in times of adversity. "If someone munches on a plant for lunch, its meristems provide backup tissue."

Whether of plant or animal origin, stem cells are, in a word, humbling. Naturalists were in awe of the fruits of their labor long before their actual identification, and even before cells in general were revealed to be the central units of plant and animal tissue in the nineteenth century. For the story of stem cells in animals is very much tied to the long history of regenerative science, and the absorbing mystery of how, for instance, a salamander can grow back its tail. Here and there in the stem cell field today, one finds a certain stigma attached to regeneration science, perhaps because, as one scientist observed to me, "it brings to mind amphibians who lose and grow back legs, and may make stem cells seem too far out there," when stem cell researchers want desperately to confer credibility on their field. And yet, the ideology behind stem cells does combine, and so interestingly, these two separate streams. On the one hand, stem cells, or regenerative cells like them, are pivotal when it comes to a salamander regrowing its missing tail—or leg, jaw, eye lens, or retina. On the other, these cells are also the reason why bone marrow transplantation has been saving human lives since the 1970s. This is not to suggest that stem cell advances will enable humans to grow a new leg anytime soon. As a sign of where things are headed, however, in 2003 tissue engineers at the University of Illinois, Chicago, used bone marrow stem cells from rats to make a replica of a human jaw.

After whole organs began to be successfully transplanted in the 1950s and '60s, there continued to be tissues in the body that were viewed as difficult, if not impossible, to repair—the brain, spinal cord, and heart, for instance. Then stem cell biology and a general turn toward regenerative medicine began warming up, and nowadays, although researchers are still searching for effective treatments for a great many ills, there isn't a single part of the body that is considered beyond medical reach. The concept of using cells as medicines, and replacing the bad with the good, has helped bring about this new confidence, and it's understandable why. Constructing a drug out of molecules that must hit an unseeable nail squarely on the head in a specific tissue is a tall order. But with cells, it's much easier to believe that they can work anywhere, because for as long as anyone remembers they have been working everywhere.

What a blockbuster assignment to make a daisy in the laboratory. How much simpler to go out and pick one.

If one could direct the fate of stem cells—that is, the specialized type of cell they become—the extrapolated mature cells could be transplanted into a problem area, where, if all went according to plan, they would overcome substandard, diseased cells and lead to healing and recovery—for example, islet cells for diabetes; photoreceptor cells for diseases of the retina; hepatocytes for liver conditions; oligodendrocytes for multiple sclerosis; spinal cord neurons for injured spinal cords; one type of muscle cell for a problem heart; another type for urinary incontinence; bone cells for skeletal disorders; or special ear-dwelling sensory cells for hearing. The list is as long as the number of specialized cells in the body that can fall prey to a disorder.

One scientist at the National Institutes of Health observed to me, "I think there's been a shift in attention to cell biology because drug companies are spending billions of dollars trying to develop things without necessarily understanding the mechanism by which a disease arises—the cell types that are involved, or what's going on with a cell. In order to develop a drug, you have to be intelligent; throwing money at something isn't going to solve it. You have to understand how it works before you can fix it."

In the late winter of '04, I received the welcome news that an old friend and a new acquaintance, both of whom had been suffering from serious illnesses, were showing signs of improvement for the very first time. Both had received newly designed cell therapies—in one case, stem cells, and in the other case, mature cells—that up until recently did not exist. Novel cell treatments like these seem to be on the increase. And yet fitting cells to the body represents such a new and complex area that it would be remarkable if unforeseen complications associated with attempts to fashion and apply cell medicines don't test scientists' wits and mettle to the utmost for a long time to come.

In the spring such a test arose. Studies, by and large, have invited guarded optimism that bone marrow stem cells can possibly help improve the condition of heart attack victims, perhaps by generating new myocytes, or muscle cells, that fortify the heart. Then came the latest news. Stem cells injected into the coronary artery of dogs led to an overproliferation of cells that clogged the small arteries. In addition, a mouse study found no signs that bone marrow stem cells, when introduced after cardiac trauma, resulted in new myocytes or heart tissue. More analysis is needed to better pin down exactly how bone marrow cells help the heart, which otherwise they do seem to do.

The lead investigator of the dog trial, Richard Vulliet at the University of California, Davis, thinks that his own negative findings "are solvable" and stands by the belief that stem cells or their descendant cells will eventually prove advantageous for human hearts in trouble. Across the stem cell field, one scientist after another voices similar sentiments. They want negative results to come forward; they want to know the worst, so that they can better understand exactly what they are up against and get to the other side. Because, as so many of these early pioneers will tell you, they are in it for the long haul.

1

Plant or Animal?

If there were no regeneration, there could be no life.
If everything regenerated there would be no death.
—Richard J. Goss, Developmental Biologist

Many Junes ago, in the early summer of 1740, a thirty-year-old tutor from Geneva—his name was Abraham Trembley—walked out into Holland's countryside to collect bits and pieces of Nature that might tease the minds of his young wards, the two sons of Count William Bentinck. Stopping by a ditch on the count's estate that connected to a stream that passed not far from the town of The Hague, the tutor selected a horsetail, a handful of duckweed, and a plump water lily. Not until he returned to the count's manor did he notice, with the help of his hand-held magnifier, the countless little green nubs clinging to the stems of these plants.

If the samplings fascinated his young pupils for more than a minute, Trembley was to be congratulated. More to the point of this story, those little green nubs would very much fascinate Trembley. In fact, the very same oddities, together with other creatures that possessed similar properties, in time would motivate an army of naturalists to probe and paw at Nature so thoroughly that entire populations of plants and animals would be at risk of being trampled. So riveting did those little green nubs prove to be that some historians would go so far as to say that they inspired the beginnings of modern biology.

Whether the tiny specimen was plant or animal, the young naturalist couldn't at once say. Under his magnifying glass the mystery object looked fairly inconsequential—not unlike a thin scrap of tubing—and offered no clues. Trained as a mathematician and an ace at problem solving, Trembley took it upon himself to try and solve the identity of the little nub, whose firm clutch on freshwater plants made him fairly certain that it was, at the very least, alive.

It was green and stationary, so it probably was a plant, he initially guessed. But the more he stared through his magnifier, the more he changed his mind. The object appeared to have something like a head, as well as skinny arms emerging from its head. Trembley would later describe these appendages in his *Mémoires* as "arms shaped like horns." And sometimes the arms—or were they tentacles?—did, in fact, move about. Or did that happen only when he inadvertently sloughed the water the object sat in? Maybe his first guess—plant—was correct after all, and the waving appendages were branches. A third possibility was that he had stumbled on an example of a "plant-animal," a life form that some people imagined must exist but no one had found. It was theorized (this representing but one theory among a chaotic panoply of theories that characterized Trembley's era) that a Chain of Being linked plants to animals, and it seemed a reasonable supposition that plant-animals, or *zoophytes*, occurred as intermediaries somewhere in between.

Plant? animal? or zoophyte? As much as Trembley kept changing his mind, he slowly became more and more convinced that his found species was an animal, largely because it moved forward by contracting and expanding "much in the same way as do inchworms," he later jotted down. Before long he would also observe that it used its spindly arms to catch water fleas, which it stuck into its extremely small mouth and appeared to eat! He would hold to this animal view, even though the more of these tiny individuals he studied, the more he saw that their arms varied in number. One individual might have five arms while another had eight, a variability seen in plants but never in animals.

One simple experiment, Trembley decided, could solve the plant-or-animal conundrum. He would cut his "insecte" (a word used for all manner of beast in those days) in two, and if it were animal, as he believed it must be, one or both halves would surely expire. But if he was wrong and it was plant, both halves would presumably stay alive and grow bigger again, just like plant cuttings.

So that's what he did. Using sharp scissors, he cut his "insecte" straight across its middle, creating an anterior piece identifiable by its head and arms, as well as a posterior piece. After putting both parts in a glass of water, he watched and waited. Because he was thinking *animal*, he imagined that while the anterior piece just might hang on to life, soon he'd see the posterior piece curl up and die. As he elaborated in his *Mémoires*, "I assumed that the second part was only a kind of tail without the organs vital to the life of an animal. I did not think that it could survive for long separated from the rest of the body."

The days passed, but as Trembley stood witness with his magnifying glass, the expected was not happening. Neither piece of his green creature was giving up the ghost, and, in fact, *both* portions kept moving about. Even more amazing, little by little both pieces began getting longer. On day 9, Trembley thought he even saw arms starting to form on the posterior piece. "I continued to see these protuberances throughout the day, and I became extremely excited and impatient for the moment when I would know clearly what they were," he wrote. By day 10 there was no question that they were arms—first five of them, then eight. Within two weeks, each of the two pieces had regrown into a whole individual!

Although it had happened right before his eyes, Trembley was incredulous, especially over the posterior piece's resiliency. "Who would have imagined that it would grow back a head!" he exclaimed. The behavior of the organism—its moving and its eating—made him continue to think *animal*, but what an unusual and remarkable animal. "I had not the least expectation of being a spectator to this marvelous kind of reproduction," he marveled.

To test the animal's rejuvenative capabilities further, Trembley began a series of experiments. Maintaining a fine balancing act, whereby he cradled an "insecte" encased in a water drop in the palm of his left hand while wielding either scissors or a boar's bristle—also used for cutting—with his right hand, all the while constantly swapping these implements for his magnifying glass, he went about cutting one green "insecte" after another either transversely or lengthwise, and in varying ways. For instance, he cut one polyp transversely into four pieces, then another longitudinally into four pieces. Low and behold, every piece became a full individual, each with a head. In another astonishing outcome, he discovered that when he sliced the head end down to nearly the animal's midpoint, each of the two longitudinal segments acquired a head. The result was a most disconcerting two-headed, Y-shaped critter. Through an elaborate process of slicing the head lengthwise, then waiting for two heads to bud back, slicing again, then waiting some more, he found he could even design a seven-headed animal!

It was at this point that he called the organism a *hydra*, after the difficult-to-kill mythical monster with nine heads. (By Edith Hamilton's account: "one of the heads was immortal and the others almost as bad, inasmuch as when Hercules chopped off one, two grew up instead.") Hydra is what this spectacularly regenerative little creature—which belongs to the phylum Cnidaria, making it a cousin to the sea anemone and the jellyfish—has been called ever since. As it happens, freshwater lakes, ponds, and streams all over the world carry dozens of species of hydra. Its saltwater equivalent, colonial hydroids, similarly festoon the oceans.

Historical records tell us that Trembley wasn't the first to describe the hydra. Among those before him was Anton van Leeuwenhoek, the Dutch naturalist renowned for beholding a wide variety of microscopic living things—including bacteria, single-celled slimed molds, tiny roundworms, and even blood cells and sperm cells—with his crude two-inch-long, hand-held microscope. When, in 1702, Leeuwenhoek plucked hydra from a stream, that stream very likely ran by his home town of Delft, Delft being situated only a few kilometers down the road from the stream that coughed up Trembley's hydra a few decades later. There is a faint chance, then, that the two men might have retrieved their hydra from the very same rambling waters.

Nor was Trembley the first naturalist to describe the process of regeneration in animals. As the ancient myth of a monster with returning heads suggests, since Aristotle's day there had been reports, both fictional and real, of animals regaining lost appendages. The tail of a lizard, the claw of a crab, the leg of a cockroach: missing parts of a not inconsequential number of critters had been observed to grow back like new. A salamander, it would be noted, could even replace the same leg many times over. Just before Trembley did his experiments, René-Antoine Ferchault de Réaumur—the esteemed leader of the Paris Academy of Sciences and himself a student of regeneration—had in fact caused quite a stir with his drawings of crustaceans and their regenerating limbs.

But although Trembley was hardly the first naturalist to be awed by the spectacle of regeneration, he was the first to conduct a systematic study of the regenerative ability of an entire animal, as opposed to just one of its parts, and the first to report some substantial findings, according to biologist Howard Lenhoff, a Trembley authority who himself spent years probing hydra at the University of California, Irvine. It astonished Trembley to discover how closely the process of regeneration followed the small-to-big growth pattern seen during an animal's development. A little piece from an older hydra could grow into the entire animal just the way an embryo did. But this was incredible! How was it possible that the head of a hydra—or the middle section of a worm, for that matter—had the power to act like an embryo?

Trembley's experiments impinged on sacred ground. To many minds, his demonstration that an animal severed in two could yield two whole animals wasn't amazing—it was unthinkable. At the time of Trembley's hydra trials, which lasted from 1740 to 1744, many natural historians and educators were hooked on certain rigid Rules of Nature derived from Calvinist or Cartesian doctrine, and one such Rule declared that the only way animals could reproduce was sexually, through a male and a female. Yet here Trembley was waving evidence of a mode of procreation that bypassed this union. Look, his experiments demanded, you can cut one of these creatures into several parts, and all parts turn into full individuals! Such was the incredulity produced by his experiments that Trembley, in a letter to Réaumur, observed of his skeptics, "Apparently these gentlemen have some cherished system they are frightened of disturbing."

The Genevan naturalist came to notice that if left to their own devices, his teensy green creatures did reproduce sexually, just like other animals. However, they also had the ability to make more copies of themselves by budding, an activity that was curiously similar to a plant's extending side shoots. Oblong buds formed roughly two-thirds of the way down a hydra's body column, and each bud eventually detached and inched off into the watery beyond to become its own little hydra. It would be shown one day that if a parent hydra budded continuously, in a month's time it could give rise to as many as twenty chips off the old block—or *clones*, individuals that are genetically identical to the parent.

As was to be expected, Trembley's evidence of asexual procreation in the animal kingdom was too iconoclastic to be believed by everyone overnight. Some of his contemporaries, unable to reconcile a hydra's incredible regenerativeness when cut in two—or four, or eight—remained convinced that his green oddity was a plant. Voltaire, for one, with an eye toward the hydra's tubular shape considered it a type of carrot or asparagus.

Many other Europeans, however, were mightily impressed with the completeness of Trembley's experiments and accepted them as valid. It helped that the Enlightenment was underway; novel observations and ideas that once might have been branded as ludicrous were now discussed with fervor. Réaumur, one of Europe's most distinguished scientists, was satisfied that Trembley's small regenerative life forms indeed were animals and spread the word that the young Genevan's investigations might bring important new knowledge of hidden life forces. Because the green nub's cylindrical body and head-circling tentacles reminded him of a miniature octopus, Réaumur referred to the animal as a "polyp."

Someone else who was fascinated by Trembley's investigations was his cousin Charles Bonnet. Ten years younger and also from Geneva, Bonnet coincidentally had encountered another form of nonsexual reproduction in Nature—parthenogenesis, which connotes "birth without father"—shortly before Trembley had launched his hydra experiments. When scrutinizing plant parasites called aphids, Bonnet noticed that the eggs of female aphids could develop and produce offspring without fertilization by male sperm.

Eager to explore the boundaries of regeneration as his cousin was doing, Bonnet chose worms to work on, and he and other naturalists would provide abundant proof that a mud worm was a master of regrowth. An inquiry he pursued much later in his life, beginning in 1779, deserves greater description, however, because by that time Bonnet was no longer focused on a small invertebrate but on a larger animal—a salamander. Like humans, salamanders are vertebrates with backbones, yet salamanders and other tailed amphibians have the distinction of being the only vertebrates that can proficiently regenerate a lost appendage during adulthood. A remarkable assertion by Lazzaro Spallanzani, another leading naturalist of the day, was that a salamander's eye, if injured, could bounce back as completely as its tail did, and Bonnet, perhaps motivated by his own increasingly poor vision, decided to look into the Italian's claim by dislodging the eye of a salamander. Here is biologist-historian Charles Dinsmore's account of Bonnet's dismal yet revealing attempt:

> His own eyes and skills, he suggested, were not up to the task and he succeeded only in badly mutilating one side of the poor animal's head. While incompletely removing the eye, he noted that the lens had popped out onto his thumb nail (his eyes could not have been too bad if he could see the diminutive lens of a salamander eye!). He set the animal aside, feeling

disturbed by the nature of the operation and thinking that the animal would be blind in one eye for the rest of its life. To his surprise, eight and a half months later the eye appeared almost normal and at 11 months, he removed the eye and determined that, among other things, the lens had indeed regenerated.

With his eyesight steadily waning, Bonnet must have viewed the salamander's restored eye with fabulous wonder, if not downright envy. Biologists would later declare that they would give their right arms to understand regeneration, and Bonnet must have felt the same way. Tantalizing questions likely brushed across his mind, such as—*Just how far do regenerative capabilities extend in more complex animals? Is there a glimmer of any such capability in humans?* If he was as astute as historians have portrayed him, the regenerative prowess seen in many lower species must have caused him to speculate that although a human could not spontaneously replace an arm, leg, or eye with a new arm, leg, or eye, perhaps scientists of the future would be able to figure out how to recreate parts of the body. At least one eighteenth-century visionary made his musings known on this subject. Voltaire was said to be so confident that humans would master the skills to regenerate themselves, he expected they'd be able to acquire whole new heads. Certainly many people could benefit from such a strategy, he intimated.

All the singular observations made by Trembley and Bonnet—as well as Réaumur, Spallanzani, Mazolleni, Müller, Guettard, Thévenot, Lyonet, De Jussieu, and other early students of regeneration—created an infectious interest in this enviable phenomenon, encouraging people to press their noses closer to Nature than ever before. One short-term consequence was that snails, particularly those in the French countryside, fell victim to an intense manhunt. Hoards of naturalists went tromping over hill and dale decapitating these gastropods in order to see if, as reported by Spallanzani, their heads really would regrow. Indeed they could, as could portions of a snail's antenna, mantle, and foot. Poor pestered snails!

Some historians maintain that it was this utter captivation with regeneration in the eighteenth century that would sweeten the way for experimental zoology and, over time, the entire modern sweep of biological sciences. To whatever extent this may be true, it stands to reason that in Trembley and Bonnet's era, when so little was known about how and why living things lived and grew and breathed and moved, an amphibian's physical capacity for regeneration must have seemed so foreign to humans as to be totally bewildering. How, for starters, could a full-grown creature regrow parts of itself when it had already undergone the process of development, from a zygote on forward? And by what means did the newly regrown material—say a crab's new claw—so seamlessly operate with the rest? Or how on earth could one section of an organism give rise to a wholly unrelated section, as in the case of the hydra's head end re-forming its foot end? Or strangest of all, how could a piece of an animal that had no mouth, hence took in no food—as was true of a hydra's foot end—re-form the creature's head end? How was this possible?

But it was too early in the history of science to begin to say. The cell—biology's coming empire—was still cloaked in inky obscurity. The answers to such questions would have to wait for tools that would take the eye much deeper into biological minutiae. Only in the previous century had Robert Hooke, with the help of his magnifying glass, glimpsed the intricate design of cells in a piece of cork, hardly guessing their central role in Life. After that, it may in fact have been Abraham Trembley who first witnessed the activity of cells dividing and multiplying. With the aid of his magnifier, he observed single-celled diatoms and stentors—aquatic organisms—dividing in two, yielding more diatoms and stentors. But he does not appear to have realized the worth of these observations, either that he was observing cells or that Life largely reaches fruition through the division and multiplication of these tiny universes.

Not until 1838 and 1839 would two Germans, Matthias Schleiden and Theodor Schwann, make perfectly clear that the cell represents the organized fabric of both plant and animal tissue, the vital component of all living things. It would be a few decades more before the notion that Life is made up of a continuing wave of cells begetting cells really took hold. Still more decades would pass before scientists gained any substantial knowledge of how cells divide, how they stack into tissues, and how they are sustained. After Schleiden and Schwann, as many as 150 years would go by before biologists determined that a hydra has ten different types of cells that are organized into two layers, and that the creature's remarkable gift of regeneration hinges on embryonic-like cells that are as ubiquitously present in hydra as oxygen molecules are in air.

In the eighteenth century, whatever lengths Trembley, Bonnet, and other naturalists went to in their endeavors to fathom how regeneration happened, they couldn't have had a clue. They didn't have the resources to get to the bottom of their questions. Not that that stopped them from coming up with some interesting theories. Bonnet was under the impression that an animal's lost part was replaced by *pre-formed* material already inside its body. This thinking grew out of the popular supposition that held that human embryos arose from minute pre-formed organs in a parent's egg or sperm. Another explanation for regeneration, espoused by a German biologist, was that an organism inherited *Bildungstrieb,* a strong inner force that was capable of reorganizing an animal's structures.

Greatly influenced by Réaumur, who avoided speculating about why things in Nature happened, Trembley offered no theories about regeneration. Unlike many of his contemporaries, including his cousin, he didn't try to fit his findings to a fashionable explanation. "Trembley didn't like theories and stayed away from them," notes Howard Lenhoff. "He felt that theories kept people from knowing what fishermen probably had observed and known about regeneration since way back in time. He felt that all that counts is your data, since interpretations will change as more is learned." Patient observation combined with smart experiments, believed Trembley, would in time reveal the workings of Nature. This "empirical" approach, which emphasizes observation over theory, would prevail as one of the Enlightenment's foremost contributions to modern science.

Trembley may have seldom wondered aloud about how a hydra performed its magic, yet he must have sometimes silently brooded over what inner properties allowed it to bounce back the way a pasture did after pruning. Perhaps in the end he simply realized that these inner things, which were undoubtedly too small to see, were meant for another generation's discovery. He and his contemporaries' dearth of knowledge about the mechanisms underlying living systems set up a roadblock beyond which they could not travel. Due to all his empirical pokings, however, Trembley would be crowned by future biologists as "the father of experimental zoology," when, scores of years later, the considerable interest in regeneration that he, Réaumur, Bonnet, and a swarm of other naturalists had stirred up would explosively reignite. Just as Trembley predicted in his *Mémoires*, in time new observations would bring long-awaited explanations:

> In order to extend our knowledge of natural history, we must put our efforts into discovering as many facts as possible. If we knew all the facts that Nature holds, we would have the explanation of them, and we would see the Whole which these assembled facts fashion. . . . Nature must be explained by Nature and not by our own views. These are too limited to envision so grand a Design in all its immensity.

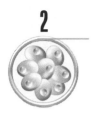

2

MOUSE STRAIN 129

The stem cell, I believe, is the interpretation of regeneration.
—John Hearn, Reproductive Biologist

Having introduced the world to the hydra's wondrous ability to regenerate itself, Abraham Trembley was brutally honest about what he had accomplished. That finding, he acknowledged, was not "the fruit of long patience and great wisdom, but a gift of chance."

Two hundred years later, the embryologist Leroy Stevens—Roy, to his family and friends—would similarly insist that his discovery of a tumor-laden mouse made shortly after he joined the staff of the Roscoe B. Jackson Memorial Laboratory in Bar Harbor, Maine, was due to luck, not genius. And what luck. The odds of bumping into a mouse strain bearing such a singular burden were infinitesimal.

He was fortunate as well, he confided to a colleague some years later, to have stumbled on this unusual sighting while working at an establishment that gave its scientists the freedom to pursue whatever they bumped into. "I felt perfectly free to do anything I wanted, and didn't have to account to anybody," remarked Stevens, a compact fellow with a long nose, dark wavy hair, a ready sense of humor, and an independent streak a mile long. "I very slowly got this thing off the ground. I mean, what do you do when you find something as *rare* as that!"

Today, it is widely held that Stevens's lucky break and all his science thereafter went a long way toward launching the field of stem cell biology. For a century logic had been whispering that cells inside an animal's early embryo must serve as the basis for all the differentiated, specialized cells that form organs and systems, and build the adult. Zoologists suspected that similar embryonic cells must lurk in the tissues of adult starfish, crabs, salamanders, leeches, and other regenerative creatures. What other explanation was there for the ability of these animals to replace a missing part out of thin air? Once his attention was given over to the strange sight he beheld in Mouse Strain 129, Stevens was able to peer into a microscope at a rodent's version of these immature, unspecialized cells. Although he never studied their potential to differentiate in depth, his research made the existence of these unique cells so noteworthy and real that scientists the world over would become mesmerized by them and hopeful that their special powers might one day revolutionize medicine.

It all began in the winter of 1953. Stevens, then a junior researcher at the Jackson Laboratory, had been assigned to look for outward differences in inbred mice that might signal inherited defects in certain genes. He was going about this task when he noticed that a chinchilla-colored male mouse in the strain he was investigating had a distinct abnormality—an enlarged scrotum. The mouse was immediately dissected and slides of its tissues examined. As a colleague confirmed, the testicular growth was a tumor that also arises in humans. It was the notorious *teratoma*—Greek for "swollen monster"—and true to its name, it could be as monstrous in appearance as it could be deadly. (For our purposes here, the term *teratoma* will be used for both varieties of the tumor, benign and malignant.)

Of the great range of tumors that strike humans, teratomas represent but a small fraction. Most are found in the gonads—usually the ovaries, where they are almost always benign. By contrast, they rarely develop in the testes, but when present in that location, they are almost always malignant and a death sentence if left untreated. Very occasional reports place them in other parts of the body: head and neck, heart, liver, stomach, chest cavity, uterus, spinal cord, and even under the eyebrow. By the early '50s, they had been spotted in only a few nonhuman animals—cow, sheep, and stallion, with some also seen in the ovaries of occasional female mice. The one beheld by Stevens was the first ever described in the testis of a male mouse.

The grisly contents of teratomas had fascinated pathologists for centuries. Unlike so many other tumors, which usually contain one type of cell, a teratoma could wind up a jambalaya of different cell types and tissues. Those that had been growing for a while and getting larger might contain muscle fiber, nerve, nodules of cartilage and bone, skin, intestinal epithelium, pigment, fat, or other cells and tissues, intermixed with fluids. Wads of hair often were the tumor's most prominent feature, while it wasn't uncommon to find a few baby teeth sitting in the midst of this strange mélange. Especially in younger tumors, undifferentiated cells—small, round, and plain—were fairly easy to spot under a microscope's lens, sometimes surrounded by mature tissue. These immature cells were stem cells—referred to in those days as "embryonic" cells. Generally, the older the tumor became, the more its stem cells proliferated and differentiated, just as happens in an organized way to immature cells inside an embryo. But in a teratoma it was unorganized, a jumble.

What was bizarre, even for a pathologist, was finding a teratoma's confusion of tissues senselessly growing where they didn't belong. What could be more disconcerting, for instance, than to find a teratoma with teeth inside it growing in a woman's ovary?—the only example of tooth development outside the mouth. A physician in the seventeenth century had reported the even stranger sight of a bone with indentations shaped like eye sockets inside the testicular teratoma of a young Frenchman. Stevens, himself, when studying teratomas from freshly killed mice would come across "twitching and pulsating movements that correspond with the histological findings of abundant striated skeletal and cardiac muscle," as he noted in one paper. Cardiac muscle cells inside a mouse's testicular teratoma pulsated in unison, just as they did in the heart.

Like others before him, Roy Stevens was instantly captivated by the teratoma. Having earned his Ph.D. in experimental embryology from the University of Rochester the previous spring, where he had been whipped into shape by the famous embryologist Hans Holtfreter, Stevens appreciated a teratoma's resemblance to an embryo, albeit a disorganized and nonsensical embryo. It, too, was a growth that most likely began as a single embryonic cell and grew larger as cells divided, proliferated, and differentiated into specialized cells of the ectoderm, mesoderm, and endoderm—the body's outer, middle, and inner leaves. Strange as it seems, some scientists saw a teratoma and its differentiating cells as an excellent chance to learn more about a mammal's embryo and its cells, and how they underwent specialization. The tumor's malignant state also provided an instructive window into abnormal development, and why some cells miss various cues and fall out of control.

Any watch post into the earliest days of a mammal's development would be a boost to embryologists. Sequestered inside the body, a mammal's developing embryo wasn't as retrievable as those of a multitude of other classes of vertebrates, whether bird or reptile or fish, and certainly not nearly as accessible as the fertilized eggs of the European newts, American frogs, and other amphibians that Stevens had studied at Rochester. Amphibians had long been embryologists' favorite friends: Their fertilized eggs, which develop outside their bodies, are easy to collect in large numbers, and the size of the eggs— four to five times larger than fertilized mammalian eggs—makes them relatively easy to examine. Since the mammalian embryo was less accessible, even by the middle of the twentieth century there weren't many mammalian embryologists. There just wasn't enough raw material to solve embryology's foremost mystery as it applied to mammals: How did a single cell—the zygote, or fertilized egg—manage the colossal feat of expanding into a highly complex individual?

The chief stages of early development in all vertebrates include *cleavage*, wherein the zygote undergoes a series of divisions that create a ball of many smaller cells; *gastrulation*, wherein the enlarging embryo develops the three germ layers (ectoderm, endoderm, and mesoderm); *organogenesis*, wherein the germ layers interact to form specific tissues and organs; and finally *birth*. In the human case, a little being comprised of trillions of superbly calibrated cells enters the world already knowing how to wail for his or her supper.

Autopsying hundreds more mice from Strain 129, Stevens came across several other males that had the same testicular growth. There weren't many of them, but he knew he was on to something. For the tumor to keep popping up, however infrequently, a defective gene or genes must be getting passed down, generation after generation.

While an inherited teratoma had never been glimpsed in a mouse, or in any other mammal for that matter, this was precisely the sort of genetic phenomenon that had compelled Clarence Cook Little, a distinguished biologist and former college president, to start the Jackson Laboratory in 1929. "Prexy" Little was a mouseologist and had been since his boyhood, when he had raised squeaking mice along with pedigreed pigeons, rabbits, and guinea pigs in his family's Brookline, Massachusetts, home. It was while he was a student at Harvard, where he had been swept up in the new science of genetics and taught by the influential William Ernest Castle, that Little began the mouse matings that would lead to the first-ever inbred mouse strain. Later, at Cold Spring Harbor Laboratory, he helped to produce others of the earliest, most enduring inbred colonies. By mating a litter's brothers and sisters over and over—for at least twenty generations—you could create a line of mice that were almost as alike genetically as identical twins, and any observable difference between two such strains of mice might reveal a genetic difference.

Little believed that inbred strains would validate his hunch that some cancers were the vicious outcome of defective genes. When Roy Stevens came on board in the summer of '52, Little was still at the helm, still directing his scientists to hunt for evidence that could demonstrate that at least some varieties of cancers arose because of the inheritance of aberrant genes, and still remarkably ahead of most other scientists in this pursuit. The idea that inherited mutations could account for cancers in families was still some years away from gaining wide acceptance.

Only one percent of Mouse Strain 129, it turned out, was afflicted by teratomas. Still, the value of this unusual inbred mouse line wasn't lost on Stevens. By examining these mice, perhaps he could figure out where in the developing embryo a teratoma started. It seemed a good bet that the pathology began before birth, because the tumor could be spotted a week after birth. Maybe he could even discover what caused the tumor, and establish some biological mechanism common to all cancers, about which so little was known.

With Dr. Little behind him, Stevens plotted out how he would locate a mouse's teratoma beginning cells. Essentially, he would inspect tissue from the testis region of younger and younger fetuses, working backward through development until he arrived at the tumor's earliest appearance. To get up to speed, he read everything he could find in the Jax Lab's library on teratomas, not without sneaking a Lucky Strike or two, which put him on thin ice with the librarian, Joan Staats. (Stevens, it seems, was a bit of a devil. Back during his stint in Officer Training School, he had graduated with the most demerits, an accomplishment he takes some pride in and wants inscribed on his tombstone, according to his daughter Anne Wheeler.)

Historically, teratomas and their weird jumble of tissues had been blamed on demons. More recently, those that grew in the ovary were thought to represent an ectopic pregnancy, which occurs when an embryo develops outside the uterus. But the theory that made the most sense to Stevens, as it did to other researchers, proposed that a teratoma formed when a primordial germ cell in the embryo—a precursor of egg and sperm cells—fell out of line. Another Jackson Lab scientist, Elizabeth Russell, had helped trace the pathway that early germs cells took when they migrate from the yolk sac into the embryo's gonadal region, where they mature into egg or sperm. With his mouse pups in mind, Stevens imagined that all it took to start a teratoma was for one migrating germ cell to go astray in the developing embryo. Although the resulting tumor usually manifested itself shortly after a mouse, or child, was born, sometimes a rogue cell sat dormant for decades, which is why teratomas could suddenly make an appearance later in life.

If Stevens was to prove that a misbehaving germ cell in an embryo initiated a teratoma, he would need plenty of mouse mothers supplying him with plenty of tumor-laden embryos. Within months, however, he ran into a serious obstacle. The laborious work of dissecting 3,557 mice and inspecting their testicular tissue had yielded only thirty teratomas. "I could have abandoned the project then and there," he told an interviewer years later. "But I was young and stubborn, and it seemed to me my work had a very real potential if I could find a way to increase the incidence of tumors dramatically and thus avoid having to raise, kill, dissect, and examine hundreds of thousands of mice in the hope of tracing the development of the tumors."

Stevens had plunged into his new project with the same dauntlessness he had shown as an Army lieutenant in World War II, when he participated in the Sicilian Campaign under General George S. Patton and served as a forward observer in reconnaissance aircraft. "Roy was a war hero," recounts Barry Pierce, a pathologist and friend of Stevens. "He was decorated several times for bravery. When he received his final medal from General Patton, Patton looked at him and said, 'You're racking up quite a collection of these, aren't you!'"

Now, more than bravery, Stevens would search for resiliency and determination to get him through a long dry spell when his science seemed to be going nowhere. After two years' time and 17,000 inspected mice, he finally hit upon a substrain in which one of every fifty males had a teratoma. In a blink, he had twice as many tumorous mouse pups to work with. But he needed still more, or the research might drag on forever. He tried feeding his mice known carcinogens, even blitzing them with radiation. But these attempts went nowhere.

His work was creeping along so slowly, he worried that the American Cancer Society would give up on him and cease its funding. To keep his ACS contacts happy, he wrote up detailed papers about Mouse Strain 129's bizarre tumors and their chaotic landscapes of undifferentiated and differentiated cells. It was "generally accepted," he frequently stated, although not yet proven, that a teratoma's growth in the testis or ovary of a developing mouse was due to "undifferentiated pluripotent cells" going astray. In his own estimation, the culprit was a germ cell in the gonads of a late embryo. Yet for all he knew, the culprit might instead be a stem cell in the early embryo.

Stevens's papers were often accompanied by photographs that depicted fields of small, round embryonic cells inside a teratoma. As the tumor grew bigger, these stem cells made more and more of themselves, along with differentiated fare. Its versatility made a stem cell a most unusual cell. Other cells upon dividing usually resulted in two equal daughter cells that resembled the mother cell. But a stem cell had choices. It, too, upon dividing could give rise to two daughter cells—stem cells, in its case. But it could also give rise to one stem cell plus a differentiated cell; or two differentiated cells.

Stevens spent more and more hours in the lab, and fewer hours at home with his wife and, by the late '50s, two young daughters and younger son. He tried to set Sundays aside, however, as a day in which he and his family could explore one scenic corner of Mount Desert or another. Having grown up in the Buffalo, New York, suburb of Kenmore, Stevens had felt cheated as a boy not to live closer to the ocean. (Coincidentally, his family had resided at #129 Lasalle Avenue, the very same number of his mouse strain.) When the Jackson Laboratory hired him, and he and his wife, Jean, came to Mount

Desert, it felt as though they had definitely arrived in Eden, the name originally given to Bar Harbor when it was settled in 1796.

Some incoming scientists discovered they weren't meant for the Maine coast, which could be gray and forlorn in winter, and quickly moved on. But not Stevens. A consummate naturalist, he would never tire of the island's rugged beauty. Anne Wheeler recounts that when she, her sister, and brother were growing up, their father's excitement over natural wonders was unfailingly contagious. "There was the time in the middle of the night," she remembers, "when the cat brought in a bat, and he woke all of us up and dissected it on the spot." Also high on her list of childhood memories are the many Sundays in summer when the Stevens and other Jackson Lab families converged on Compass Harbor with their picnics. There, after throwing their sandwich crusts to the resident gulls George and Martha, the embryologist and his children would spend blissful hours wading in the cove's immense tidal pool, hunting for crawly, slippery creatures hidden under rocks and seaweed, Stevens as content as a clam in mud.

While savoring the island's spectacles, this observant scientist would feel the tug of a connection between the plant and animal species around him and the featureless embryonic cells—stem cells—that he witnessed scattered in the midst of the teratomas he was extracting from mice. It had been quietly appreciated for some time that Nature's fecundity must have lots to do with the embryonic cells that underlie the cellular construction of plants and animals. These young cells were the road to cellular diversity, the fountain through which every kind of specialized cell flowed. From stem cells, as well as progenitor cells—cells that were a touch more differentiated but still had stemness—evolved myriad lineages of maturing cells that diverged down further branching avenues of form, function, nuance, and color, culminating in the creative detail and specialization that allow creatures to thrive on land, in water, or in air. An osprey's talon or a wolf spider's eight eyes, wild indigo's bright flower or the light bones of a ruby-throated hummingbird or the

nectar the hummer sips on—every bit of biological variety was descended from stem cells. Everything had to come from something,
and that something was the stem cell.

In the mouse teratomas that Stevens examined in the lab, the
more advanced specimens could contain as many as fourteen different mature cell types—bone, intestine, blood, hair, muscle, cartilage, nerve, and more—the whole lot likely springing from a single
cell. The proficiency of a stem cell, whether from a tumor of the
gonads or an early embryo, was as mystifying as a wizard's ability to
change willy-nilly into anything he desires to be. It had something to
do, Stevens knew, with the genes inside a stem cell's nucleus and
their getting switched on and off. But knowing this didn't make a
stem cell seem any less magical.

Many of the marine creatures that Roy Stevens and his children
spied in the tidal pool at Compass Harbor—starfish, crabs, sea
anemones, jellyfish, sea urchins, and others—were among the very
critters that had been the objects of unending experiments in the late
nineteenth century and had gone a long way toward enlightening
naturalists about embryonic cells and their distinctive capabilities.
By the year 1900, cells were a fairly new addition to science's Book of
Knowledge. Only in 1838 and 1839 had they been identified as the
vital "pores" of living tissue. Even less time had passed since, in 1855,
Rudolf Virchow had observed that "every cell comes from a cell,"
which had led to the realization that the microscopic world was teem-
ing with dividing cells, multiplying cells, and differentiating cells.

In the 1890s, experiments with sea urchins brought the great
surprise that cells in the early embryo were far more *plastic*—that is,
flexible—than previously imagined. The German biologist Hans
Driesch had expected to find that when he separated an urchin's

beginning two-cell embryo, each cell would develop into only the part of the animal it was fated to become. So the work of other scientists had indicated. "But things turned out as they were bound to do, and not as I had expected," Driesch later confessed. He was dumbfounded to discover that each of the two cells had the capacity to develop into a complete larva, a complete organism. It happened again when he broke apart a four-cell embryo: an entire larva ensued from each of the four cells. The larvae were smaller than usual but normal nonetheless.

The fertilized egg obviously possessed a power that enabled it to give rise to a complete individual, but now Driesch was showing that even after it had divided once, then once again, its daughter cells also had this *prospektiv Potenz*—"prospective potency," as he called it. Their fate was far from sealed, apparently, although the older the embryo, the more the potency of its cells seemed to diminish. The early embryo, Driesch declared, was endowed with "harmonious equipotential," and perhaps as a direct result, the terms *totipotent cell* and *pluripotent cell* gained regular use by the early 1900s. A totipotent cell can generate every cell connected to an organism; a pluripotent cell, nearly every cell. A stem cell bearing such potency was described by other adjectives as well, including *indifferent* or *undetermined*, both words referring to a cell's undifferentiated state. As for the term *stem cell, stammzelle* appeared in the German literature of the nineteenth century, although its English-language equivalent did not cross many lips until Roy Stevens's era. Even up through the 1970s, researchers were more apt to refer to stem cells as "embryonic" or "embryonal" cells, according to scientists of that generation.

Driesch had presciently stated, "Every cell, during development, carries the totality of all primordia." But he and his contemporaries fell short of knowing *why* a cell had such potential. Although Gregor Mendel had theorized in 1865 that minuscule entities inside pea plants were the messengers of inherited characteristics, it would be well into the next century before the genes that lay tucked away in a cell's nucleus would be fully valued as the blueprint for inherited traits, as well as the means by which a cell knew to specialize down a certain lineage pathway, be it blood or bone.

During the tail end of the nineteenth century, meanwhile, interest in the phenomenal regenerative skill that abided in so many grown creatures was booming. A burgeoning fleet of American and European naturalists, among them Driesch and other revered experimental biologists of the day, were sinking their teeth into the subject, hoping to solve the countless questions that had stymied Trembley, Bonnet, Réaumur, and other earlier regeneration pioneers. Why had Nature given this remarkable ability to smaller creatures but not to more complex animals? You only had to get up to birds on the evolutionary ladder, and the talent was practically gone. Some smaller creatures, however, were as regeneratively lackluster as humans. Regeneration appeared to be a quirky talent that was distributed unevenly across the animal kingdom.

Could it be, these scientists asked themselves, that embryonic cells endowed with *Potenz*, the likes of which Driesch had caught sight of in the early embryo, resided in the tissues of grown animals as well? What better explanation could there possibly be for the miraculous regrowth of, for instance, a lobster's claw? Yet when a naturalist took to his microscope to find these powerful cells, he didn't necessarily see them. In the late 1800s, the biologist Francis Herrick, fascinated by how a grown lobster could drop a claw and handily replace it with a new one, methodically scrutinized a lobster's anatomy for a "store of embryonic cells" that might explain this trick. His search proved fruitless, and as he noted in his monograph *The American Lobster* (1895), "The examination of serial sections through this part of the limb reveals nothing but normal tissue cells. Embryonic cells may be present but are not discernible."

The emerging consensus was that the animal kingdom exhibited two varieties of regeneration—the ordinary kind and the extraordinary kind. The former consisted of the routine repair and replacement of tissue that every organism underwent. The continuous replenishment of skin, for example, was as important to a person's everyday survival as a tree's ability to grow back its leaves.

But it was the second and far less frequent type of regeneration that was more interesting to regeneration scientists, particularly since humans lacked it—an animal's expedient restoration of an entire missing part. This capability is termed *epimorphic regeneration*, referring to the remodeling of multiple tissues. By the early twentieth century, biologists perceived that a "bud," or group of cells, forms at the boundary of the lost part, and that these cells rapidly proliferate and differentiate into the range of tissues necessary for remaking the structure. Salamanders and other tailed amphibians were observed to be especially able regenerators. Lose a leg, a tail, a jaw, or an eye lens? A salamander could grow a replacement in a matter of weeks or months. Impressive, too, were the American lobster and numerous crab species that could so deftly regenerate a claw lost to battle or caught under a rock. How many more pickpockets there might be, biologists joked, if they too could drop an arm and scamper off, knowing they'd soon have another?

(There's the somewhat parallel situation of a kelp crab's clever means of escape, as pointed out to me by Robin Cooper, a biologist at the University of Kentucky. If a seal tries to eat a kelp crab, the crab will give the seal a good pinch, dropping its claw while doing so. Which is why every once in a great while a seal will be spotted with a crab's claw dangling from its nose.)

At field stations from Wood's Hole to Pacific Grove, regeneration scientists struggled to divine information about the restorative prowess of earthworms, silkworms, flatworms, fiddler and gulf-weed crabs, hydra and various fish, leeches, frog larvae, lizards, salamanders, and Mediterranean medusae (jellyfish). If you left a hydra in darkness as opposed to sunlight for twenty-five days, did regeneration proceed more slowly? If you starved a salamander, did that affect its limb restoration? Was there a relationship between a crustacean's molting period and regeneration? How many arms could you remove from a Brittle-Star and still see recovery? Darkness didn't appear to markedly slow regenerative growth; nor did starvation, although the restored legs of starved salamanders were quite a bit more spindly than those of well-fed salamanders, as the biologist Thomas Hunt Morgan assiduously observed in a 1906 study. Molting and regeneration did seem to share anatomical features. And, as might be expected, the Brittle-Star didn't survive the removal of all five of its legs. Poor starfish.

As in Trembley's day, most early-twentieth-century reports fell short of deep explanation. But while regeneration would have to wait for the Molecular Age for further unraveling, some tangible observations were pocketed. Zoologists agreed that for regeneration to take place in either grown plants or animals, sufficient embryonic cells must be present. But where did they come from? One possibility cited by Eugen Korschelt in his tome *Regeneration and Transplantation* (1927) was that cells "left over from embryonic development" still lingered in the adult body, "very insignificant complexes of indifferent cells which we cannot perceive." A second possibility noted by Korschelt was that mature cells at the site of a missing appendage first *de*differentiated back into an unspecialized embryonic state, and then, upon regaining totipotency, they and their cell offspring could proliferate and differentiate anew into the sundry cell types that were needed to reform an entire appendage from scratch.

This alleged retro ability was reported as far back as the mid-1800s, shortly after the cell theory put cells on the map as the common ingredient of all living matter. "Some facts concerning the regressive differentiation of cells have long been familiar," noted zoologist Henry Wilson in 1911. He refers to an 1856 journal report that states that a sponge's choanocytes, mature flagellated cells, could "de-specialize" and possibly make "a complete return to the indifferent (totipotent) state." Wilson himself imparts that the activity of straining small pieces of a mature sponge through "fine bolting cloth such as is used for tow nets" produced immature cells that proliferated and then *re*differentiated into mature sponge tissue. There might be "other animals besides sponges and hydroids in which the somatic cells when forcibly disjoined, will fuse and give rise to totipotent regenerative material," Wilson elsewhere comments.

Not everyone believed that mature cells could dedifferentiate. Korschelt mentions that there were those who stuck to the opinion "that once cells have developed along specific lines, they are in no case able later to change tack." Yet Korschelt felt that due to the abundance of evidence, "there can scarcely be any denying the possibility that development processes are reversible." Immature cells could mature, and so apparently could mature cells regress, he believed.

Day in and day out Roy Stevens was tipping the lid on stem cells. They lay just below his nose, bathed in his microscope's circle of light. Granted, the stem cells he examined were from mouse tumors. Most probably had malignant properties and were a shade different from normal stem cells in embryos. Nevertheless they were stem cells, and for anyone interested in investigating the far reaches of their capacity to differentiate into a range of mature cells, Stevens's Mouse Strain 129 and its teratomas made them available. Normal stem cells had not yet been isolated from embryos, not even a mouse's; so the tumor kind were most unique and alluring. Another scientist might have put everything aside to concentrate on these special cells and their talents, but Stevens's sole bent was to understand how a teratoma began and grew, in hopes of "gaining control of the disease in humans," he would explain. So focused was he on digging down to the roots of this cancer, that when evaluating him for a possible fellowship, one Jackson Lab colleague observed that if Stevens had any shortcomings, they arose "from his singlemindedness" at the bench. Continued this observer, "I think he has chosen an independent course and will continue so."

To make progress, Stevens needed still more tumor-beset animals. In 1958 he gambled that among the Jackson Laboratory's considerable stocks of mice, he would find a mutant mouse that, when mated with his 129 strain, would increase the incidence of tumors. It was a shot in the dark. The hoped-for mouse had to have a specific genetic susceptibility for teratomas. A long three years and eight generations of mice later, Stevens met with success. Thirty percent of his mouse offspring were now heritably doomed to teratomas, and he could finally move on to a full-blown investigation.

If the tumor was caused by a deviant germ cell, he guessed he would catch the evidence in a fetus's genital ridge, the tiny region that develops into the testis or ovary. He and his technician Donald Varnum painstakingly worked backward through mouse fetal time: gestation day 19 (birth), day 18, day 17, and so on. After endless dissections, in 1964 they at last came upon the first sign of tumor growth—on day 12. There, in the genital ridge, a germ cell that should have become a sperm cell started erroneously growing like an embryo, its cells dividing and multiplying. "By tracing the testicular teratoma back to its precise origin, we had focused on the very beginning of a cancerous process for the first time in history!" Stevens later exulted to a Jackson Lab chronicler. The journey, which had taken twelve long years, helped prove Prexy Little's premise: Inherited mutations in genes could indeed bring about a cancer.

Now that he had the tumor on its genetic knees, Stevens wanted to design drugs to combat it. By the late '60s, he could perpetuate a near limitless supply of tumors by transplanting a fetal mouse's tiny genital region into the testis of an adult mouse. Teratomas resulted because the genital ridge's germ cells, its stem cells, raced into chaos when buried in unfamiliar territory. Then Stevens did an experiment that established once and for all that an errant germ cell commenced a teratoma. He transplanted fetal genital ridges that contained almost no germ cells into the testes of adult mice—and no teratomas bloomed. What stronger proof was needed that errant germ cells caused mouse teratomas? Case closed!

An experimental embryologist to the hilt, he next grafted mouse embryos into odd places in adult mice—the kidney, for instance—and embryos frequently turned into disorganized teratomas. This was further confirmation that stem cells, in this case those in an embryo, didn't like being plunged into alien surroundings. As witnessed here, put an embryo and its stem cells where they didn't belong and a teratoma ensued. "This showed how easily an embryonic cell and a cancer cell could move one to another. There was a certain similarity between these two cells," notes Davor Solter, who in 1970 was in a laboratory in Yugoslavia that made the same finding.

As Virginia Papaioannou, a developmental biologist at Columbia University, describes, "Both an embryo and a teratoma are made up of stem cells that divide rapidly. The main difference is that stem cells in the embryo have controls put on them at certain points—they know when to stop dividing." But if they are placed outside the uterus, "those growth controls fail, leaving a tumor to develop."

Stevens was convinced that, in a reverse situation, if one put a teratoma's stem cells into the uterus where they belonged, they would switch from malignant to normal. Don Varnum recalls Stevens's conviction: "'Don,' Roy would say to me, 'if we could get some of those cells in the right environment, I think they would participate in normal development.'" Adds Varnum, "Well we tried, but we couldn't get it to work."

Despite his many trials and setbacks, Stevens nevertheless felt that luck had constantly visited him. And in the early '70s, it did so again when a Jackson Lab colleague, Seldon Bernstein, encountered and brought to Stevens's attention an inbred mouse strain in which roughly fifty percent of newborn females had ovarian teratomas by the age of three months. The strain and its aberration were "a great shot in the arm for me," Stevens later recalled, for he would discover that the ovarian tumor's cell of origin, just like that of its counterpart in the testis, was a deviating germ cell that began growing in the fetus as if it were fertilized. This false embryo "developed beautifully for quite a while," Stevens described. "It looked just like a normal embryo, but then it would all get mixed up. It became disorganized and you had a tumor instead of a baby mouse."

In the 1970s, Roy Stevens's mouse-derived teratomas and their stem cells would begin attracting worldwide attention. But before they did, there was only one other scientist on the planet who was as fixated on teratomas as Stevens, and our story at this point would be incomplete without mention of Gordon Barry Pierce and the way he so productively straddled pathology and embryology. By "the late 1950s and early 1960s, Roy or I formed a rather small, but intensely interested audience whenever the other gave a paper," Pierce, a luminous and loquacious pathologist, later related. "To most oncologists, we were the fellows with 'the funny little tumors.'" If people laughed at them, it wasn't for long. Together Stevens and Pierce would become the famed pioneers of teratoma biology.

Pierce's immersion in teratomas started in 1953, the very year that Stevens happened onto his first teratoma-burdened mouse. During his pathology residency at the University of Alberta Hospital, the Canadian had been in the impossible position of trying to stave off a testicular tumor in a three-year-old boy. Chemotherapy was not yet an effective option, and the drugs that would one day save children from this deadly tumor were still years away. When the boy died, Pierce was enormously sad and also frustrated by how little was known about testicular tumors. Determined to render them less mystifying and less lethal, he began requesting teratoma tissue from Roy Stevens in the late '50s, and due to their mutual interest in teratomas, the two men quickly became close friends.

By '64, Pierce had wrung some crucial findings from teratomas. He and Lewis Kleinsmith, a medical student of Pierce's at the University of Michigan, Ann Arbor, proved what had only been postulated before, that a teratoma's assortment of specialized cells definitely arose from one pluripotent stem cell. (Stevens, at the time, was pinning down the actual type of pluripotent cell involved: a germ cell.) Kleinsmith and Pierce had removed a single cell from a teratoma, a cell they imagined was pluripotent, and implanted it in the abdominal cavity of a mouse, and the implanted cell had duly produced a teratoma with a teratoma's typical bedlam of cells and tissues.

Working with Frank Dixon at the University of Pittsburgh, Pierce had also shown the extent of a teratoma's malignancy, and their revelations proved stunning. The widespread belief was that every cell in a malignant teratoma must be malignant, both its stem cells and its mature cells. Yet Pierce and Dixon uncovered that the tumor's malignant state was due entirely to its stem cells, for as long as they remained undifferentiated. When the stem cells differentiated, they actually stopped being cancerous. Pathologists already had a name for a teratoma's undifferentiated stem cells: embryonal carcinoma cells, or EC cells.

The great surprise that a teratoma's cancer appeared to lie in its stem cells, and not its mature cells, produced an uproar among oncologists. The notion that malignant cells could differentiate into benign cells "took dogma right by the throat," says Pierce, because it ran counter to the conviction that *once a cancer cell, always a cancer cell.* A cancer cell was only supposed to "divide and divide; that's part of the ritual of being malignant." It wasn't supposed to be able to change to a benign state, or become anything other than what it was.

Today, Pierce speaks glowingly of the experiments Armin Braun performed at the Rockefeller Institute (now Rockefeller University) in the early '60s that also left convincing evidence that tumor cells could become normal cells. What might seem curious, yet shouldn't given all that plants and animals share at the molecular level, is that plants can also be afflicted with a teratoma. Braun used the cells of teratomas from tobacco plants affected by crown gall disease—these plant tumors being as much of a jumble of immature and mature cells and tissues as an animal teratoma—and he showed that tumorous buds, if grafted to normal plant tissue, could regain normalcy.

Braun recognized that genes within the cells were behind this changeover. "The cells were made tumorous simply by activation of some of the ordinarily inactive genes; the cells returned to the normal state when those genes were again repressed and rendered nonfunctional," he observed in a *Scientific American* article.

Pierce concluded in the early '70s that malignant stem cells were but "caricatures" of normal stem cells and originated when normal stem cells slipped out of their regular routine. The observation was not a new one. "But Barry is the one who really provided the experimental basis" for what he referred to as a stem cell theory of cancer, says Ralph Parchment, a pharmacologist at Wayne State University and former student of Pierce's. At the core of the theory: If too many undifferentiated cells form and not enough of them mature, their nonstop self-renewal could lead to cancer's destructive growth. Pierce's observations prompted him to recommend a treatment measure. Rather than attempt to kill cancer cells through such toxic means as chemotherapy, he entreated, why not try to control their rampant division by getting them to differentiate into benign tissue?

To this day, Pierce stands adamantly behind this idea. However, the evidence for "differentiation therapy"—the practice of attempting to cure cancer by getting its immature cells to mature—is far from conclusive. One of several questions posed by skeptics is, if a cancer stem cell can be changed into a benign state, say by treating it with a certain chemical, would its genes really and truly be normal?

A remarkable experiment in the mid-'70s nonetheless provided confirmation, in the opinion of many scientists, that cancerous stem cells indeed could be nudged into a normal state. The question posed by three separate groups was, if you place malignant EC cells from mouse teratomas in the early embryo, will this normal environment impose the right controls and induce EC cells to differentiate into benign, normal cells and tissues? Roy Stevens, as we saw, often said as much to his assistant Don Varnum: "Don, if we could put these cancerous stem cells into a normal embryo, I swear they'd go back to being normal." Upon injecting a teratoma's stem cells into early mouse embryos, two American teams and one British team basically got the same results. Most of the time these implanted cells contributed to the development of normal adults; their malignancy swung toward normalcy. The groups that independently published this evidence were led by Ralph Brinster at the University of Pennsylvania, Beatrice Mintz at the Institute for Cancer Research, and Richard Gardner at Oxford University.

With the blossoming of the above experiment, a vision was coming alive. What if you could alter the genes of stem cells in ways that diseases alter genes, and then return the modified stem cells to the embryo? As long as the modification ended up in the developing embryo's egg and sperm cells, you might be able to create entire lines of changed mice that mimic a specific human disease. These transgenic animals—animals that have been permanently changed in this fashion—might be invaluable for studying all sorts of diseases. Since normal embryonic stem cells were not yet obtainable from mice, a teratoma's malignant stem cells were at the center of this vision.

It all hinged, however, on getting the gene changes into an animal's germ line—the lineage of egg and sperm cells that carry the genetic code of one generation over to another with ceaseless continuation. If successful, scientists would then know how to modify Life.

These days, as long as a child's or adult's teratoma is discovered in time, this cancerous tumor is virtually curable through the administration of cisplatin, a chemotherapy drug. Cisplatin is one of a trio of drugs that helped restore bicyclist Lance Armstrong to health when he was diagnosed in 1996 with a testicular tumor that contained a small element of teratoma.

Cisplatin and other chemotherapy drugs arrived irrespective of the ground covered by Roy Stevens and Barry Pierce. And yet these famed teratoma pioneers produced breakthrough observations that would impact the far corners of biology, and therefore the study of disease and medicine as well. Barry Pierce, who is now retired and spends much of his time attending to his bonsai collection and working at his bench—his woodworking bench—alerted many scientists to the possibility that some cancers might be a stem cell disease. Today, more than ever, cancer researchers are investigating this claim.

Roy Stevens, who is also retired, still lives within an arm's length of the ocean. Because of his lucky sighting back in '53 of a most unusual mouse, biology's horizons would expand in every direction. Researchers throughout the world would request his famous mouse, eager to study its tumors, their precocious little cells, and the rainbow of mature cells derived from them. Even a tumor's stem cells had lots to say, and embryologists, cell biologists, geneticists, pathologists, oncologists, and other investigators would lean in to listen.

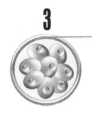

3

The Purple Cell

There is a lot of biology between the stem cell and mature cell.

—Harvey Patt, Radiobiologist

In the 1950s, as Roy Stevens was trying to wring more teratomas from his chinchilla and albino mice, opportunities for scientists to study stem cells under the lens of a microscope were next to nonexistent. Teratomas were a rarity in laboratory mice. Normal stem cells, meanwhile, had yet to be retrieved from animal tissue. Embryologists were well aware that the normal kind briefly resided in the recesses of the early embryo, and that other cells with stemness came and went in the older embryo and fetus. Yet if scientists were going to unbury these master cells, they had more to learn about where and when they occurred in early development and how long their potency in fact lasted.

Signs abounded that regenerative cells were also present in the tissues of adult animals. Clearly, in humans they lived deep in the skin; how else did skin remain a constant when cells on its surface were constantly dying? Bone and the intestine manifested a similar turnover of cells. But since these and other renewable tissues were largely composed of mature cells, where did any intermingling stem cells lurk? Overwhelmed by futility, microscopists scanning tissue for a stem cell would soon give up, for in a crowd of cells one cell blends into another. If truly present, these special cells, along with

47

being featureless, were either very reclusive or scarce. Imagine flying over the Pacific and looking down upon wave after wave in an attempt to find a featureless wave!

Nowhere in an adult was a stem cell more anticipated than in the hematopoietic, or blood-forming system. The vision of a hematopoietic stem cell, in fact, had been planted in people's minds back in 1909 by Alexander Maximow, an eminent St. Petersburg embryologist. And there is reason to believe that earlier scientists had planted the same vision in Maximow's mind. A mouse fetus harbored a blood-forming stem cell-like cell in its liver, the fetal liver being one of the first places that blood is made in a developing mouse. But if blood stem cells were also present in an adult animal, they hadn't shown themselves. Or at least none had been reported.

Starting in the '50s, a long line of experiments—two of which will be described shortly—projected an increasingly real sense that a blood stem cell did exist in adults, for what else could account for the constant replacement of blood cells, which, just like skin cells, were largely short-lived? Even though scientists couldn't see this reclusive stem cell, it was imagined to reside inside the marrow-filled cavities of numerous bones of the skeleton, where it continuously churned out new blood cells that joined others in the bloodstream by way of the blood vessels that intersect bone. This persistent stem cell—or three separate stem cells, as some believed—appeared to be the source of replenishment for the blood's three major lineages of cells: red blood cells (erythrocytes), the oxygen carriers; white blood cells (leukocytes), the infection fighters; and platelets (thrombocytes), the facilitators of blood-clotting.

We know better today what brief lives blood cells have. From birth to death, platelets and most white cells last only a few hours or days. Red cells are a bit hardier, surviving for 120 days or so. Given that some 25 trillion red blood cells are flying through the human bloodstream at any one time, the blood-forming system has an impressive turnover rate to maintain. With every passing second roughly two million red cells are expiring, and two million others are hurtling out of the bone marrow and into circulation to take their place.

The blood's ability to renew itself was one sizable clue back in the 1950s that stem cells must lie concealed in bone marrow. There was also another clue. Mature cells throughout the body were thought to be created in one of two ways—either when stem cells or progenitor cells divide and produce descendants, or when mature cells divide and produce more of themselves. In the case of red blood cells, they could only stem from stem cells (or progenitors). That's because by the time they reach maturity they lack a nucleus, and therefore cannot divide to make more of themselves.

But it wasn't only the blood system's normal workings that indicated the presence of stem cells; so, too, did assaults on the bone marrow's formation of blood. It had been known for several decades that ionizing radiation, the type emitted by gamma rays, could decrease the bone marrow's production of blood cells. It wasn't altogether clear why this happened, although some scientists theorized that radiation wreaked havoc with marrow-residing stem cells. When hit by radiation, the marrow basically stopped making and spewing out blood cells. The horrific casualties and radiation sickness that resulted from the American atomic bombs dropped on Hiroshima and Nagasaki had produced plentiful evidence that rapidly dividing cells in the intestine, skin, and blood—which were thought to be stem cells—were terrifically sensitive to radiation.

Hard evidence of this, according to Irving Weissman, a cancer biologist at Stanford University, came from experiments done by radiation biologists in the late '40s and early '50s, when they exposed mice to the effects of increasing amounts. "If they gave 200 rads, that wasn't enough to kill mice," Weissman relates. "With 700 rads, some might die, depending on the strain. At 800, most would die; 900, all would die. Those that died didn't have enough platelets to clot their blood; enough red blood cells to carry oxygen; enough white blood cells to fight off infection."

Scientists had cleverly decided to turn radiation's ill effects to medicine's advantage. Why not aim its lethal rays at the fast-dividing cells of a cancer? It didn't take long for this notion, once tested, to pay off. During the '50s and '60s, for instance, cells affected by Hodgkin's disease were found to be exquisitely susceptible to radiation, and Hodgkin's went from being an always-lethal disease to sixty-to-seventy percent curable. Even though radiation was known to also initiate cancer, the probability of this occurring appeared to be quite low, with the benefits of radiation therapy thought to far outweigh its risks.

Once they got more adept at using radiation, medical researchers began testing an idea that was plausible yet speculative. If you gave an adult mouse a lethal dose of radiation that wiped out its blood-making system—resulting in a pathetic mouse that was bound to die—would bone marrow from a healthy mouse that was infused into the ailing mouse's bloodstream get the ailing mouse's blood system back on track? Procedurally, it was a bit of a leap. "Who would believe that if you put cells that are normally in bone marrow into a place that you don't normally find them—the bloodstream—that they'd find their way back" to the bone marrow, points out Irving Weissman. But apparently they could and they did. The first effective animal-to-animal bone marrow transplants were reported in the early '50s. "It was considered quite miraculous at the time," remembers James Till, a biophysicist and senior scientist at the Ontario Cancer Institute in Toronto. "In effect, a bone marrow transplant given after an otherwise lethal dose of radiation rescued an animal from a sentence of death."

Despite Professor Maximow's earlier pitch that the bone marrow was home to stem cells that spun out the mature cells found in the blood, the prevailing view in the '50s was that some sort of chemical "factor" released by the transplanted marrow is what got blood cells in the depleted animal humming again. But in 1956, a British team led by Charles Ford, a biologist at the Atomic Energy Research Establishment just south of Oxford in Harwell, England, produced proof that the magic stimulus wasn't a chemical; instead, it was the presence of certain cells in the donor's marrow that as much as breathed life back into a transplant recipient's blood-making system. Recollects Weissman, "It went from being a factor to cells. But then the question followed—*what cells do it?*"

Early in the century, Maximow had been ridiculed for his vision of a blood stem cell. But now people wondered if his prophecy might not be true. Somehow the professor's theory had to be tested, although if a blood stem cell was as reclusive as it seemed to be, it would be like setting a trap for a ghost. This brings us to the first experiment. Performed at the Jackson Laboratory in the late 1950s, it didn't attract much attention in the larger world. Yet what a fortuitous little experiment it would prove to be in terms of the scientific advances that would come on its heels.

Located up the hill from Roy Stevens's bench space, and sharing the same view of Dorr and Cadillac Mountains, was a lab run by Elizabeth Russell, a cell biologist and geneticist who was loved and admired as simply "Tibby." Russell stood a solid five-foot-two, every inch packed with strut and principle, especially when it came to science. A mentor to younger researchers, she would leave behind in people's memories volumes of rich remembrances—about how she orchestrated the Jackson Lab's rebuilding of its mouse colonies after Bar Harbor's windswept fire of '47 in which all but a thousand of the laboratory's 100,000 mice had perished; about how, early on, she appreciated the mouse's value as a living stage for the study of genes and proteins associated with human diseases; about how, in collaboration with Beatrice Mintz, she had tracked a germ cell's migration into the developing embryo, which had helped Roy Stevens to trace the misbehaving stem cell that started a teratoma growing in a fetal mouse.

Not much slipped by Tibby Russell, attests Jane Barker, a Jackson Lab hematologist and coworker. Even when she nodded off at meetings, which she did without fail, she always awoke in time "to ask the best question in the room," recalls Barker.

Russell's early aspiration as a scientist had been to decipher genes connected to an animal's pigmentation. This had led her to study abnormal pigmentation and its link to anemia, a red blood cell deficiency, which further had led her to explore the genes implicated in anemia. To assist her, she had at her disposal a strain of white mice with jet-black eyes that she and various colleagues had spent years breeding and studying. The mice had an aberrant gene that gave rise to numerous deficiencies, not the least of which was a severe form of anemia.

As the '50s progressed, the prospect of transplanting bone marrow from animal to animal, donor to recipient, had gotten Tibby Russell thinking. Thus far, the animals at the center of these experiments were normal—normal, that is, until researchers made them infirm by casting radiation over their entire bodies, which affected their bone marrow and incapacitated their blood-making ability. It followed that an animal's immune system was suppressed, since it could not generate a full quota of white blood cells. If the donor and recipient were not genetic kin, quieting the recipient's immune system vouchsafed that the transferred bone marrow cells wouldn't be rejected.

Russell's idea was instead to transplant blood-forming cells into mice that actually had a medical problem. She had in mind her lab's very anemic mice. "We could have given them blood transfusions," recounts Seldon Bernstein, then an Atomic Energy Commission Fellow from Brown University whom Russell had recruited because he knew his way around the hematopoietic system. "But blood transfusions are ephemeral. Mature red cells have a finite life span and don't last very long. The question was, could we give these anemic mice more permanent cells—stem cells that could keep replicating?" While a blood transfusion worked off *mature*, nondividing cells in circulating blood, blood-rendering stem cells from a different source might refurbish, or even cure, the anemic animals.

Bernstein, together with Lois Smith, a coworker, had worked out a system for retrieving blood-forming cells from the tiny liver of a mouse fetus. Given that they were the predecessors of cells that would populate the blood's different lineages, today they would be looked upon as stem cells or progenitors, although back then Bernstein remembers referring to them as "embryonic cells." It was their fairly immature status that made these fetal cells such an attractive choice for transplantation. Other labs were showing that immature cells ran a smaller chance of rejection by the immune system than more mature cells, their very featurelessness possibly allowing them to sneak by the transplant recipient's immune system undetected. "At the time, nearly fifty years ago," points out Bernstein, "we were fumbling at the edge of the known—and speculating" about the compatibility of tissues taken from different individuals. Even though the donor and recipient mice belonged to the same strain, and their tissues likely were compatible, using embryonic tissue seemed safest. "The idea was to give the graft every opportunity to survive and replicate by eliminating genetic differences," he observes.

Russell's experiment lifted off in 1956. The first task was to obtain sufficient numbers of cells from the minuscule livers of fifteen-day-old mouse fetuses that were "smaller than the half-moon of your littlest fingernail," notes Bernstein. This fussy work was accomplished by pulling apart the tiny liver with needles, carefully avoiding injury to the cells. Bernstein describes it as a bit like trying to pick apart the seeds of a pomegranate without mashing them. He and Russell then put the cells into fluid and injected the whole lot into the tail veins of dozens of their anemic white black-eyeds and sat back to wait and see if the transplanted cells engrafted—lived and divided and created gallons of new red cells that boosted the anemic mice's badly starved blood.

Their patience wasn't wasted. Within sixty days a large majority of the treated mice showed signs of being permanently cured. Some would live for as long as 600 days and might have survived even longer had they not been euthanized so that their blood cells could be inspected. (Healthy mice of the same strain normally live about 700 days.) What truly thrilled Russell and Bernstein was that the transplanted cells seemed to really fix the blood-forming system of their mice, and in no superficial way. Evidence of the mice's prior anemia—too few red blood cells that were over-sized—had completely disappeared, and now their marrow was producing regular numbers of normal-sized red cells.

"It was the first rescue for anemia. It was the first time a stem cell transplant had been used to cure disease in an experimental animal," observes Jane Barker at the Jackson Lab. A landmark, to be sure.

Russell and Bernstein went on to correct several other types of genetically induced anemia through an assortment of methods. Meanwhile, their earlier experiment's demonstration that immature, or stem cell-like, cells of one mouse could repair another mouse's damaged blood signaled that it was the eve of something very substantial and new in medicine. As their published report in 1959 had concluded, if ways of avoiding tissue rejection could be identified, "it may be possible to use cell implantation generally for the therapy of genetic defects" associated with red blood cells.

The era of cell therapy was definitely dawning.

If bone marrow from one grown animal could rekindle the blood of another, presumably stem cells resided in the marrow of adult animals. But still it was the same old story. The wait continued for someone to come along and pluck out a blood stem cell and demonstrate: *Here's a blood stem cell and watch what it can do.* "The hypothesis was that stem cells existed. But no one knew for sure that they existed, and no one had been able to recognize them," recollects

Ernest McCulloch, a hematologist and senior scientist at the Ontario Cancer Institute. Logic favored this cell's existence; few disputed that. But there was "a big controversy in hematology" over a related issue, recounts McCulloch. "It was between people who thought that each one of the [blood] lineages had its own stem cell, as opposed to the view that a single stem cell could give rise to several lineages." In other words, did red cells, white cells, and platelets arise from different stem cells? Or were all three produced by one and the same stem cell?

"The controversy always struck me as rather foolish," says McCulloch, "because if you push things back far enough in development, you always come to just one cell." One cell that must head the blood's different lineages of cells. Somewhat similarly, go back far enough in development and you come to one cell—the zygote— that starts the whole ball rolling. (An obvious parallel is how different animal species may have evolved from one common ancestor; and, to push back further still, plants and animals may have arisen from the same little microbe.)

The research that Ernest McCulloch embarked on in 1959 would eventually silence this controversy. Thirty-three at the time, this Toronto native's forceful intelligence left a lasting impression on people. Like his father and two uncles before him, McCulloch had become a doctor, in his case settling into clinical work at the Toronto General Hospital. When he was recruited to join the Biological Research Division of Toronto's newly opened Ontario Cancer Institute (OCI) in 1957, he saw it as a superb opportunity. "The old philosophy was to make research with your hands—everything you needed," recalls McCulloch. "But at the new institute, it was very clear that the people in charge wanted to make life easy for you. They would supply me with new research facilities—also collaborations that I wouldn't have had otherwise."

McCulloch's chief research interest was leukemia, and like any researcher interested in leukemia in those days, he wanted to explore the sensitivity of leukemic cells to various chemotherapy drugs, these drugs appearing to work because they killed blood stem cells that drove leukemia. To do this, McCulloch needed to be able to grow blood cells in culture, and first prove that the cells in his dish really were blood cells. Stains were used for recognizing dried cells on slides, but they didn't identify cells in culture, in a dish. In short, McCulloch came to realize that a good proof of blood cells, particlarly its stem cells, would be to inject them into animals that had received radiation, and should the animals bounce back, he would know that blood stem cells were part of his injected mix. This might prove a good new way, in general, of identifying blood cells and the stem cells among them. With the ability to identify these cells, you might even be able to improve ways of therapeutically targeting them, through chemotherapy, radiation, or other means.

Close to the time when Tibby Russell and Seldon Bernstein's promising mouse cure was published in '59, McCulloch was preparing a test he hoped would yield a true picture of radiation's effect on living cells. He had discussed the project with another junior scientist—James Till, an OCI physicist skilled in administering radiation—as McCulloch's supervisors had made it quite clear that they didn't want him sallying forth on his own with this particular project. "Dr. Harold Johns, head of physics and famous for his knowledge of radiation measurement, insisted that whenever a biologist such as me was presumptive enough to try and use radiation, a physicist had to be associated so there'd be no horrible mistakes," relates McCulloch. Till was equally curious about radiation's impact on cells and glad to lend a helping hand.

McCulloch's immediate aim was to investigate a hard-to-believe claim. As he notes, "There was a widespread belief that a tumor's cells were more susceptible to radiation than normal cells." Researchers assumed that this was why radiation was good at treating certain cancers; it "melted" the tumor, yet did less damage to the surrounding tissue. But along came a scientist from Colorado, Theodore Puck, who surprised everyone by finding contradictory evidence. When he exposed malignant and normal cells to radiation, both reacted with similar sensitivity. If Puck's evidence held up, it would force a re-thinking of why radiation was so effective against cancer. There was reason to mistrust Puck's findings, however. The cells he experimented with had been sitting in the unnatural circumstances of culture—in a nutrient bath in a dish—and not in a living being.

So the question arose, was Puck's finding accurate? Was all tissue, whether or not from a tumor, similarly sensitive to radiation? Or were Puck's results skewed? "One of the things Ernest and I agreed upon was what an interesting question that was. Maybe it would prove to be a myth that some tumors were more sensitive to radiation than others," recounts Till. Till might never have known the fascinating weight of such a question were it not for his parents' appreciation of education. Born and raised on the northern prairies of western Canada, he had intended to become a grain and livestock farmer like his father, but thanks to his parents' encouragement, he had ended up not with a herd of cattle but with a Ph.D. in biophysics from Yale.

To test Puck's claim, the hematologist and the biophysicist would make use of dozens of irradiated mice—mice whose bone marrow had been intentionally devastated by radiation, mice who would die within thirty days if they didn't receive fresh bone marrow. Till, the irradiator, spent a fair amount of time in a room in OCI's basement, where the radiation source was located—a lump of cobalt-60 that was about the same size and shape as a stack of quarters. Since this natural emitter of gamma rays couldn't be shut off, to restrict its rays it "was embedded in the rim of a circular block of lead, which was itself embedded in a bigger block of lead," describes Till. The entire three-foot-wide lead encasement was suspended from an arm attached to the ceiling. (At the time, due to its ability to penetrate deep-seated solid tumors in most any organ, cobalt-60 was used extensively for cancer therapy as well as animal research. The "cobalt bomb," as the therapy was called, had been developed by none other than OCI's formidable head of physics, Dr. Johns.)

To turn the source on, further relates Till, "we rotated the circular block. Then we placed a pie-shaped compartment underneath the source, each wedge holding a mouse, and rotated the mouse container so each mouse got the same dose."

With the ultimate goal of testing what radiation did to a cell, in the first leg of their experiment the two scientists injected different amounts of marrow cells into their radiation-exposed mice, which allowed them to determine the minimum number of transplanted cells it took to rescue a mouse. Next, they again infused bone marrow into irradiated mice, but this time they also irradiated the bone marrow they were injecting. Then, once again they set about determining how few cells it took to rescue a mouse. By comparing this second round of results with the first round, they expected to discover whether the infused normal cells enabled the animal to recuperate any better than the cells that had been x-rayed.

One Sunday afternoon in April of 1960, in the midst of his and Till's investigation, McCulloch drove his broken-down Dodge over to the lab to keep up with the task of euthanizing mice and removing their femur bones and spleens. An inspection of both told him how well an animal's marrow and blood cells were growing. (Unlike in humans, a rodent's spleen plays a major role in blood formation.) Up until then, he and Till had waited several weeks after a transplant before doing this examination. But on that Sunday—only ten days out from the last round of transplants—McCulloch decided to sacrifice a few animals earlier this time, curious to see to what extent their cells were engrafting. Upon cutting into the flank of the very first animal, which put the spleen in full view, right away he was struck by the spleen's unusual appearance: It was inundated with largish lumps.

McCulloch was immediately aware of what probably had occurred. Some of the marrow cells that he and Till had injected into the bloodstream of their mice must have lodged in the spleen. Each lump was far too big to be just one cell, and therefore possibly represented a miniature mountain of cells.

"When I saw those lumps, it occurred to me at once that I might be looking at clones," he recalls, or a colony of cells derived from a single cell—a remarkably prolific cell. He dissected four or five other animals that afternoon. The spleen of each one revealed the same large nodules. The thought that the spleen-ensnared cells that appeared to generate thousands of progeny might be stem cells didn't occur to him immediately. That idea would settle in some weeks later, after he and Till intellectually raked over this strange sighting in the spleen with Arthur Ham, OCI's biology director. Only then would it become obvious to McCulloch and Till that they might have inadvertently netted the blood's previously elusive stem cell.

"I probably went home and had a relaxing Sunday evening dinner with my wife and thought nothing about it, except that I'd talk to Jim Till first thing in the morning," recounts McCulloch.

Remembers Till about the next morning: "My memory is of us meeting in the hall actually, and Ernest was waving this piece of graph paper and saying, 'Look at this!'" Through careful plotting, McCulloch had determined that a relationship existed between the number of marrow cells injected into various animals and the number of bumps on their spleens. Amid every few thousand injected marrow cells, his graph suggested, there was a cell capable of grandfathering a lump's worth of other cells. Why hadn't they seen these lumps when they'd sacrificed animals at later intervals? Because by then, offers McCulloch, most lumps would have diminished, their multitudes of cells having matured and dispersed.

The two men finished the experiment that had led them to all of this. Puck had been right; normal cells and tumor cells were both sensitive to radiation. Because of Puck's finding, there'd be no dramatic changes in practice, says McCulloch, although gradually, over time, researchers would update their protocols for using X rays to treat cancer.

After finalizing that work, the Torontonians turned their attention back to their possible evidence of the long-anticipated blood stem cell. Inspecting the nodules that popped up in mouse spleens, McCulloch discovered that those that had enough time to grow contained a mixed bag of blood cells: red blood cells, certain white cells, and huge megakaryocytes that were precursors of platelets—just the sort of heterogeneous family you'd expect a multipotent blood stem cell in an adult animal to produce if given the time. But how would they possibly prove that all the cells in each nodule were truly derived from a solitary multipotent stem cell? The problem threatened to upend them.

A graduate student of Till's took on this challenge for his Ph.D. project. The student, Andrew Becker, had learned a useful nugget from Charles Ford (the same aforementioned Ford from England) when Ford had dropped in to speak to the Ontario Cancer Institute's graduate students and faculty. Ford said that x-raying cells sometimes left a distinguishable mark on the cells' chromosomes, and that this mark could be followed as a marker, since it got handed down to a cell's offspring. Doing a swan's dive off this insight, in 1962 Becker produced an elegant set of experiments. He demonstrated that when cells having this marker were transplanted into a mouse's bloodstream, most of the cells in the resulting spleen colonies bore this same aberration, proving that a lump's thousands of cells did indeed arise from one cell. (Even though red blood cells have no nucleus, and therefore no chromosomes, the marker was detected in their precursor cells.)

By now, McCulloch and Till were quite sure they had captured the blood stem cell—or something close to it. When visitors to their lab peered through a microscope to see this rare cell that hid in the marrow of an adult mouse and, quite likely, an adult human, too, Jim Till would rapidly inform them—"Oh, it's really easy to spot, because it's purple." Such was Till's standing joke. In reality the stem cell they had trapped was impossible to distinguish from other cells. "Jim knew he could get away with saying it was purple, because he knew that no one could disprove him," gruffly laughs McCulloch. "Because you couldn't see it or find it." Spleen colonies were filled with hundreds of thousands of red and white blood cells, and a featureless stem cell easily got lost amid the mountain of cells that it gave rise to.

"You have to realize what exponential growth is like," explains McCulloch. "If you had one dollar and you doubled it twenty times, do you know how much you'd have? A million dollars! Going from one cell to a million cells in twenty doublings, it would be next to impossible to find your original dollar—or cell—among the million." In short, even though McCulloch and Till had clear evidence of a blood stem cell in rodents, the cell itself remained as elusive as ever.

Based on their trappings, however, they could at least propose a definition that might help others to recognize a stem cell—any stem cell at all, whether from an embryo or a fetus or an adult. Elucidates Till, "We felt a stem cell should be defined by its potential for extensive proliferation. If it can't produce a lot of progeny, it isn't a stem cell." As for its progeny, "A stem cell should be capable of producing new stem cells like itself; otherwise, it would be on a dead-end street. And it should also have the potential to produce different kinds of differentiated cells."

Most scientists today would say that McCulloch and Till's definition has stood the test of time. Merely having a definition as a guide to this cell would speed the science along, for how can you find anything—especially an immature, nondescript cell—if you don't know what you're looking for? Andy Becker would uncover something else about a stem cell to add to this definition. Even though it could be highly proliferative, it appeared to also be capable of existing in a state of arrest, which suggested that the adult stem cells that lurked in the tissues of organs possibly lasted the entire life of an animal.

The Canadians' research validated the blood's hierarchy: Blood stem cells made progenitor cells that made mature cells. Prior to the Toronto team's findings, any knowledge of the differentiation process that produces the blood's specialized cells had been based "on

appearances," McCulloch notes. Stains could reveal differences between cell types, but how those differences arose was anybody's guess.
The Canadians' research therefore provided a quantitative method
for tracking differentiation: If you transfused *this* many cells into a
mouse's bloodstream, in ten to fourteen days you got *this* many
spleen colonies, which meant that *this* many stem cells must have
been present in the transferred blood. Creating spleen colonies became a tool, for now hematologists could analyze a mouse's blood
stem cell's range of progeny to their hearts' content.

To biologists everywhere, with the blood system hoisted high as
a model, it suddenly seemed twice as likely that other tissues in adult
mammals—the large intestine, for instance, which constantly
sloughed and renewed its cells—must also harbor rare populations
of stem cells. Aided by improving stains and markers, by the mid-to-
late '70s, other investigators would make progress pinning down evidence of stem cells in the liver, the testis, the lung, the digestive tract,
and elsewhere, and find that mammals, in general, turned over vast
quantities of cells every day, every month, every year. (The current
belief is that an average-weight adult human possibly sheds ten bil-
lion older cells every day and generates as many new ones.)

Like all cells, save for egg and sperm, a stem cell was known to
divide via mitosis, a process that splits the cell and its genetic material in two, each half becoming a new cell. Yet in thinking about how
stem cells generate differentiated cells, McCulloch, Till, and other
scientists found themselves up against some hard questions. For starters, what told a stem cell to divide? And how did it know to create
either more of itself or more specialized cells? "The language we seem
to fall into suggests the cell has a choice, that it can either divide and
give rise to two daughters, or it can choose to differentiate and cease
to be a stem cell," McCulloch notes. Yet the language we apply "may
not be totally appropriate," he advises. "Suppose a stem cell divides
and gives rise to two new stem cells, and in just a few minutes one of
those cells decides to differentiate. It just might happen like that."

One might have thought that once Till and McCulloch had

pocketed evidence of a mouse's blood-forming stem cell in 1960, it wouldn't be long before the cell itself would be isolated. (Their report was published in 1961, the same year the first human, the Soviet cosmonaut Yuri Gagarian, traveled into space.) But more than twenty-five years would pass before Irving Weissman's Stanford lab would bring this distinctive stem cell, or a relative thereof, to the surface in both mouse and human. Weissman and his team achieved this milestone by working backward from mature blood cells to an early precursor that consistently sired a full gamut of blood-cell progeny—lymphocytes, monocytes, granulocytes, and red blood cells. Praised by many, their detective work depended heavily on monoclonal antibodies—proteins that can distinguish between subsets of cells—as well as Till and McCulloch's colony assay technique.

Later still, Weissman would find that hematopoietic stem cells in adult mice are rare and yet not so rare. Only one in every 15,000 bone marrow cells is a blood stem cell; yet that amounts to as many as 150,000 such cells in the bone marrow of an adult mouse. His lab would further determine that between 20,000 and 100,000 of these stem cells pass through a mouse's bloodstream every day.

In general, the early animal-to-animal and human-to-human bone marrow transplants attempted by researchers went abysmally. Ernest McCulloch and colleagues, for instance, tried transplanting marrow into leukemia patients in the early '60s, only to meet up with the same obstacles that other researchers were experiencing.

Probably the most bedeviling problem was graft rejection, which made allogeneic transplantation quite impossible. (In an allogeneic transplant, a patient receives bone marrow from a donor, whereas in an autologous procedure, a patient receives back his or her own bone marrow, which has previously been removed and stored.) But as time marched on, dedicated researchers such as E. Donnall Thomas and George Santos in the United States, and Dirk van Bekkum and George Mathé in Europe, began to surmount the various problems, including tissue rejection. At the crux of this problem are human leukocyte antigens (HLA), proteins on the surface of cells that alert the immune system to the presence of foreign cells such as unwanted bacteria and viruses. The valuable insight that scientists learned by the late '60s was that if an allogeneic transplant was to work, the HLA antigens of donated cells had better be identical to, or closely similar to, the recipient's HLA profile.

Because of McCulloch and Till's adventure in the lab, there was no longer any question that if infused bone marrow saved the life of a patient, it was due to the graft's stem cells. Found among the millions of cells injected into a recipient's bloodstream, the stem cells home to the marrow, where they quickly proliferate and roll out their specialized blood descendants. McCulloch and Till's mouse studies indicated that, in addition, transplanted tissue can benefit from viable stem cells that remain in the recipient's bone marrow. Most will have been knocked out by radiation and/or chemotherapy, but some can be rejuvenated by the incoming flow of transferred cells.

A new, more efficient method for harvesting hematopoietic stem cells has come into prominence the last couple of years. While stem cells are still drawn from bone marrow, more frequently these days they are harvested directly from a patient's bloodstream (the peripheral blood). A person is given injections of factors that stimulate the marrow to produce greater than normal amounts of stem cells that in turn spill into the bloodstream. His or her blood is then collected, its stem cells put aside for transplantation, while other components are returned to the bloodstream. In 2002, an estimated 30,000 autologous transplants and 15,000 allogeneic transplants utilized hematopoietic stem cells, according to the International Bone Marrow

Transplant Registry, with peripheral blood stem cell grafts the predominant type of graft used in adults.

The many patients whose lives have been prolonged and saved by a bone marrow transplant are gradually catching up to what researchers have known for some time—that they owe their lives not to bone marrow per se, but to a powerful "purple" cell found within.

4

Mystery in a Dish

If you can't grow something, you can't improve its growth, because there's nowhere to start.
—Howard Green, Cell Biologist

Developmental biologists, at least those interested in mammalian development, will tell you that the early 1970s were an exceptionally stimulating time, and all because of the gift that Roy Stevens had given their field. The mice of Strain 129 and their bizarre yet fascinating teratomas were a hot item, and a small but fervent coterie of aspiring postdocs and rising and already-risen embryologists wanted this rare mouse in their labs. They delighted in the chance to study the stem cells inside a mouse's gonadal tumor and watch them grow and differentiate into all types of tissue. This preoccupation was possible because scientists were learning a new trick, namely how to keep stem cells from mouse teratomas alive in a petri dish.

"I was doing it for the very reason that so many other people were doing it," describes John Gearhart, who in the early '70s was a postdoc at what now is the Fox Chase Cancer Center in Philadelphia. "Finally we had an experimental system in mammalian embryology where we had a pluripotent cell sitting in a dish, and we could say, look, we're getting nerve cells and muscle cells and others. Thanks to what Barry Pierce did, we knew that a single cell was responsible for forming all of these other cell types."

Working at MIT was a cell biologist M.D. who found the prospect of exploring mouse teratomas so alluring that at age forty-six he had actually switched fields for this very purpose. He was Howard Green, another gifted Canadian-born scientist. Shortly before joining MIT's biology department in 1970, while still a professor at New York University School of Medicine, he had invented a novel method for assigning a human gene to its respective chromosome, a practice seized upon by medical geneticists. Yet despite this achievement, "I thought it would be more interesting to study cell differentiation in a teratoma," Green recalls, even though he "had no fixed idea of how to proceed." A cordial, trim, neatly attired gentleman who prefers not to retire quite yet, Green presently holds the position of George Higginson Professor of Cell Biology at Harvard Medical School.

Because Howard Green was the sort of scientist who liked to directly connect with other scientists whose work intrigued him, one cold day in the late fall of 1972 he and James Rheinwald, one of his graduate students, made the long drive north to the Jackson Laboratory on Mount Desert to pay Roy Stevens a visit. They soon returned to MIT with a few of Stevens's popular tumor-bearing mice. While Green had no precise plan for his research, he was curious to see for himself the assortment of cell types that a teratoma might yield. It wasn't long after Rheinwald had sacrificed a mouse, minced its tumor, and set the cells growing in a culture dish that one certain cell claimed Green's attention, because its identity stumped him. He was familiar with many types of cells, but he'd never seen the likes of this one before. It was flat and compact and nothing like a fibroblast (the predominant cell in connective tissue) with its irregular shape, or a red blood cell with its doughnut-like concave center.

Rheinwald tried separating cells of this type away from other teratoma cells, but they wouldn't grow alone. They seemed to need the support of other cells to flourish. Biologists had become wise to the fact that for cells to multiply in the artificial setting of a culture dish half as well as they did in the body, they often required the proximity of feeder cells, cells that lend support to other cells. One had only to appreciate how densely packed the body is with cells to realize that it wasn't natural for one type of cell to grow on its own, in obscurity, without neighbors. So Rheinwald thawed and made use of 3T3 feeder cells, a cell line that Green had developed in the early '60s while at NYU. (When a cell and its progeny are made to divide and therefore "grow" indefinitely, the result is a cell line, an invaluable resource that provides researchers with a near limitless and accessible source of cells.) These 3T3 cells were fibroblasts from the connective tissue of fetal mice, and labs clear around the globe had come to depend on them because of how they boosted the growth of other cells. Green's nickname for them—3T3s—referred to a necessary requirement for their growth. They had to be transferred to new flasks three times a week at a density of about 300,000 cells per dish.

When added to other cells, 3T3s first were exposed to X rays to prevent their own growth in culture—that is, in a dish filled with specially made media. This done, Rheinwald commingled the feeder cells with the teratoma's mystery cell—and the strategy worked. The nameless cell began to flourish, indicating that the added feeder cells released a protein that the mystery cell found nourishing.

Now that their cell was growing so nicely, dividing and proliferating happily, Green and Rheinwald had a much better chance of learning its identity. They already had discussed its resemblance to an *epithelial* cell. Found everywhere in the body and assuming many different shapes and structures, epithelial cells share the common motif that they form surfaces, whether the body's outer surfaces or the interior surfaces of numerous cavities and passageways. Flat, cuboid, or cylindrical, they pave the surfaces of tongue, cheek, esophagus, intestine, urinary tract, the lung's alveoli, and the ear and nose canals, to name only a few of the places they reside. To compensate for the constant wear and tear they take, Nature has made these cells highly regenerative.

Their cell's epithelial nature really spoke to Rheinwald. After getting his bachelor's degree from the University of Illinois, and before coming to Harvard Medical School, he had spent a year at the University of Wisconsin tending fast-dividing fibroblast cells, which had left him puzzling over why epithelial cells failed to similarly thrive when put in the same basic medium. With Green's approval, Rheinwald decided to devote his doctoral thesis to their unidentified epithelial cell, and maybe this time around he would determine why its growth requirements were different from a fibroblast's. Cells, scientists were realizing, were as finicky as people; culture conditions that suited one type of cell didn't necessarily suit another.

The question that Green and Rheinwald were left with was, which of the many classes of epithelium did their mystery cell belong to? One spring day in 1971, some unexpected assistance arrived in the form of Burton Goldberg, a colleague of Green's from NYU. Glancing through a microscope at colonies of their cells, Goldberg exclaimed, "Why, that's a beautiful stratified squamous epithelium!" The kind of epithelium he was referring to forms the outer barrier layer of the skin (the epidermis), the cornea, the mouth, and the esophagus. An exceedingly durable tissue, it is composed of keratinocytes, cells that synthesize a tough protein called keratin, and it was the lavish layering of stratified keratinocytes that Goldberg's eagle eyes had spotted.

Progress! But now Green and Rheinwald were left to wonder which bodily surface their keratinocyte cells actually belonged to. Skin? Cornea? Mouth? Esophagus? It remained to be seen whether keratinocytes from these areas, when placed in the company of 3T3 feeders, would grow as well as their teratoma-derived cells. They decided to first test keratinocytes from skin, which were easily obtained. Rheinwald simply contacted his wife's obstetrician at the Boston Lying-In Hospital, who set aside the circumcised foreskin of a newborn, and during a free moment Rheinwald picked up the sample, which lay ensconced in a plastic tube inside a tin can, and transported it back to MIT on the subway—not without a few thoughts about how odd a biologist's life sometimes could be. In the lab, he set about disassociating the skin cells and placed them in a dish together with Green's hearty 3T3 cells, eager to see whether they would survive and multiply. The year was 1974, and the day, Rheinwald recalls, was September 10—Howard Green's birthday.

To the best of Green's knowledge, no one had ever gotten human keratinocytes to grow outside the body, in culture, in such a way that permitted them to appreciably proliferate. Biologists had been trying to cultivate skin for a century or more, but no one had hit upon the right method. For that matter, it was difficult to convince almost any brand of cell to divide and grow in a culture dish, and that was because no one was altogether sure of what to feed cells—what culture conditions appealed to them. Once in a petri dish, more often than not cells languished and died, halting any further investigation into their behavior. Fibroblasts and a few other connective tissue cells were among the few exceptions. Less finicky than other cells, they were easily maintained on a standard fare of vitamins and calf serum.

The lure of being able to culture epidermis wasn't hard to understand. What a fabulous resource surgeons would have if they could take, for instance, a little bit of a burn victim's skin and multiply it many times over, until there was enough to cover the injured area. To heal, a skin wound needs protective covering. But in the case of many burns, a wound can't heal because epidermal cells at the wound site are destroyed, and the dermis layer below the hair follicles is unable to reconstitute the epidermis. For over a century, epidermal skin grafts had performed a terrific service in this regard. So remarkably, a person's epidermis could withstand being detached and reattached to another location on his or her body. The skin graft had to come from the same person (an autograft), because otherwise skin taken from a donor would be rejected by the receiver's immune system without fail. That left the door open for a major dilemma. If someone was badly burned over a large area, they might not have enough healthy skin left for grafting.

There was a secret to a successful skin graft, and over time that secret had been revealed to be stem cells. When removing a piece of skin, a surgeon had to make sure to shave off the section deeply enough to include the bottommost layer of the epidermis—its basal cells—along with some underlying connective tissue. Otherwise, when the piece was transplanted, like a sapling that fails to establish its root system, it might not take hold. The basal cell layer was important to include, surgeons had discovered, because it contained stem cells that enable the skin to regenerate. They constantly churn out new cells that progress through the outer layers of the epidermis toward the surface, die, cornify, and are shed, with other cells rising to follow. Human skin completes this bottom-to-top, birth-to-death cycle within three to four weeks.

"At that time, no one in the profession thought in actual terms of stem cells," explains Green. "But they knew that if they made a good graft, the grafted cells would provide progeny for the life of a person, so that implied stem cells." In keeping with McCulloch and Till's definition of a stem cell, basal cells deep down in the epidermis produced lots of progeny and enjoyed long survival. Just like stem cells from bone marrow, they had permanency.

Historians claim that the first person to successfully graft animal tissue was none other than Abraham Trembley, that wonderman of hydra research. An experiment that Trembley undertook in 1742 presented him with the strange sight of the tissues of two hydra growing together, and meanwhile shows us what deft hands and instinct he must have possessed.

Somehow overcoming a hydra's smallness—most are no longer than one-third of an inch—and drawing upon his knowledge that a hydra is essentially a hollow tube, he began his experiment by turning one of his tiny green creatures inside out like a sock. But this sleuth's imagination was only just warming up, for he then strung this inverted hydra onto a boar's bristle and proceeded to push it into another hydra's mouth, positioning it all the way down inside this second hydra. A few days later he saw something quite remarkable, which was that inner tissue of the outer animal had fused with outer tissue of the inner animal.

"I cannot explain what became of the body of the inner polyp, whether it was dissolved in the stomach of the outer polyp, or whether the inner body merged with the outer," he later wrote in his memoir. "I can definitely state, however, that I could still see the body of the inner polyp lying inside the outer polyp several days after it had been inserted. As for the inner polyp's head, I was certain that it had fused with that of the outer polyp. The lips of the outer polyp adhered to the neck of the inner polyp, and after a certain time the two heads of these polyps formed but one which had two rows of arms."

Ever since, grafting tissue has imposed a menacing complication. Unless a donor's tissue is compatible with the recipient, it can easily be rejected, its cells facing attack—in the case of humans—by lymphocytes dispatched by the recipient's immune system. Trembley didn't run into this problem in the above experiment, because the two hydra he used were from the same species, their tissues a near match. But had they been from different species, in all likelihood their tissues would not have fused. Hydra don't have lymphocytes, but "they certainly have some active form of an immune system," maintains Hans Bode, a biologist at the University of California, Irvine, who has been contentedly studying hydra for over thirty years.

As twentieth-century biologists have come to appreciate, a hydra is replete with active stem cells. Its body column consists of two one-cell-thick layers of epithelial cells—the ectoderm, which lines the animal's outer surface, and the endoderm, which lines the gut cavity—and the cells of both layers behave like stem cells, according to Hans Bode. As these stem cells do their thing, which includes continuously dividing while sometimes carrying on a function like digestion (unlike stem cells in more advanced creatures, whose sole function appears to be producing progeny cells), they move outward, not unlike the way stem cells in the basal layer of human skin do. "The geometry is a bit different, but the concept similar," shares Bode. "There's a steady state situation in which cells produced in the middle of the animal are displaced toward the extremities. Those in the upper body column are displaced into the head and out to the tip of the tentacles where they are sloughed. Those in the lower body column get displaced into the foot and sloughed at its base."

Biologists would take up the idea—how could they not?—that a hydra's wealth of stem cells accounts for its phenomenal ability to regenerate. Its capacity to regrow and remodel itself, they would submit, requires more than just stem cells; other internal forces, for instance, organize the new tissue within the context of the whole. On the other hand, if it weren't for a hydra's multitude of stem cells and their tireless yield of new cells, the hydra would be at a loss for new tissue and wouldn't be the superb regenerator that it is. Notes Bode, given Trembley's discoveries about hydra, "Trembley's discovery *was* regeneration." Which goes to say that, even though he would never know one, his discovery was the stem cell as well.

In the fall of 1974, Green and Rheinwald were unknowingly poised to extend the practice of grafting human skin beyond its current limitations.

Once they had put cells they knew to be foreskin keratinocytes into a dish with 3T3 fibroblast feeder cells, they didn't have to hold their breath for long. After a few days they observed that their keratinocytes, nourished by the 3T3s, were growing "amazingly well," describes Green. "They grew better than we hoped." Now all was revealed. The mystery cell that had stolen their attention just happened to be the special cell that serves as the regenerative basis for human skin, and the 3T3s they had used to support its growth happened to be the perfect companion, for in the real life setting fibroblast cells in the connective tissue that underlies the basal layer stimulate the proliferation of its cells. Without the 3T3s, their skin keratinocytes "wouldn't have grown at all!" notes Green.

To suddenly realize that they had crossed River Impossible and had skin cells flourishing in the foreign environment of a dish—the surprise couldn't have been greater. "It was a happy event," says Green thriftily. A happy accident as well, he acknowledges. Here people had been attempting to cultivate skin cells for ages, and he and Rheinwald had done it without any intention of doing so.

The MIT biologists would find that as long as they used their 3T3 fibroblasts as nursemaids, they could grow keratinocytes not only of the skin, but those found in every other squamous epithelium tissue. Pretty soon they were propagating keratinocytes taken from the human cornea, the pharynx, the vagina, the esophagus, and all regions of the mouth. Their laboratory hummed with incubators filled with this cell—a type of human cell, remember, that never before had been made to proliferate outside the body to any great extent.

Keratinocytes from skin remained the researchers' main interest. Certain proteins known as "growth factors," they found, quickened the cells' proliferation rate. Still, they had trouble keeping the cells multiplying, because they tended to differentiate into mature cells and stratify. Bit by bit, however, the investigators devised ways of improving their culturing technique. By transferring the cells to fresh flasks more frequently, and applying an even more judicious choice of growth factors, they were able to maintain most cells in a proliferative state, which resulted in enormous numbers of cell progeny in a surprisingly short time.

So successful were they in fine-tuning their methods, Green and Rheinwald were able to expand a piece of epidermis the size of a postage stamp to more than 5,000 times its original size. This was roughly the surface area of an entire adult human being, or about twenty-one square feet. Growing up this quantity of cells took approximately three weeks, the numbers of keratinocytes in culture doubling every seventeen hours. As Green observed, the culture conditions that his lab had worked hard to optimize enabled epidermal cells to grow faster than they did in the body.

It became increasingly clear that the essence of the scientists' culture method was its preservation of keratinocytes that had a strong degree of stemness. The procedure of transferring cells to new flasks in a timely fashion perpetuated small keratinocytes that were as strikingly proliferative as the large cells in their culture were unproliferative, and the scientists were all the more certain that the small cells they were urging to grow were equal in behavior to the stem cells in the basal layer of the epidermis that perpetuate skin renewal. "We knew that the small cells in our flasks were proliferative—that they were basal cells," notes Rheinwald, who is currently the associate director of the Harvard Skin Disease Research Center at Brigham and Women's Hospital.

Today, these regenerative skin dwellers are regarded as either adult stem cells or their more differentiated kin—progenitor cells. They are less flexible than an embryo's stem cells, but they have enough stemness to produce vast numbers of epidermal progeny. A recent finding raises the possibility that epidermal keratinocytes could possibly harbor a latent ability to produce other types of cells. As shown by Michele De Luca and Graziella Pellegrini in Rome, keratinocytes that form the eye's conjunctiva, the membrane that lines the eyelid, give rise not only to other keratinocytes but also to the conjunctiva's mucus-secreting globet cell. Previously, these two ocular cell types were thought to originate through separate lineages.

Howard Green couldn't help but realize what his lab's revelation might mean for surgeons who were in constant need of epidermal cells, whether for the sake of covering burns, or replacing ulcerated or scarred skin, or skin ravaged by a blistering disease. "We immediately knew that we had something that would make possible the cultivation of this cell type in a way far superior to anything that had existed before," Green recounts. But if push came to shove, could their flask-grown epidermis be converted to a useful form of therapy? With patients in mind, he set about designing the following protocol:

- grow up the keratinocytes
- disassociate them with an enzyme and transfer them to new flasks
- grow the cells up again and detach the entire sheet of cells from the bottom of the glassware with another enzyme
- wash the sheet to get rid of any traces of enzyme
- gingerly staple the sheet to a thin petrolatum gauze
- place the skin with its gauze backing onto a prepared wound and suture in place
- remove the gauze a week or so later

Green first applied this skin-growing technique to a living animal in 1980. After preparing a sheet of cultured human keratinocytes, he and a postdoc used it to cover a small wound on the back of a "nude" mouse, handling the tissue ever so carefully since it was as thin and friable as wet Kleenex. The mouse came from a strain that not only was furless, but lacked an effective immune system. Thus it wouldn't reject the human graft, which in the weeks ahead took hold nicely. The mouse's skin, once healed, was seen to be made up of both the graft's human epidermis as well as mouse epidermis, which had grown in from areas bordering the grafts.

Not long after that, Green tried his new technology on people. He and a colleague prepared cultures and clipped them to gauze for Dr. Nicholas O'Connor, a prominent burn surgeon at the Peter Bent Brigham Hospital, who sutured these epidermal sheets onto the small burns of several patients. Nearly half of the cultures engrafted and contributed to the skin's complete recovery.

But Green's culturing strategy would receive a far, far greater test in 1983. One day Green, who had moved on to Harvard Medical School, received a call from John Remensynder, a surgeon at the Shriners Burns Institute. Three young boys, Remensynder related, had sustained severe burns over more than ninety-five percent of their bodies. They had been painting, and as they cleaned themselves off with a solvent, one of them had struck a match to light a cigarette. One boy had died soon after the accident. The other two, brothers ages five and six years old, were expected to expire as well, since as a rule children burned that badly couldn't survive. Dressings that included strips of cadaver skin and sheets of collagen would temporarily assuage their wounds, but sooner or later their bodies would reject these foreign materials. What both boys needed for engraftment, but didn't have enough to spare, was their own healthy skin. Remensynder wondered if Green would be willing to lend a helping hand by providing the burn surgeons in charge—Nicholas O'Connor and Gregory Gallico—with his innovative skin-growing strategem.

"We were a basic research lab, not a hospital lab, and were poorly equipped to do anything on that scale," recalls Green. "It would take lots of cultures and lots of people to prepare the cultures and transfer them from our lab on Longwood Avenue across town to Shriners." Yet he knew full well that if the attempt wasn't made, the boys didn't stand a chance.

And so Green, his five postdoctoral fellows, and two technicians rolled into action. As it would turn out, they would culture 146 grafts for one brother, and as many as 233 grafts for the other brother, whose wounds proved more problematic. Each graft measured approximately seven square inches. When early-morning surgery was called for, Green and his team would arrive in the lab by 5 a.m. in order to give themselves enough time to staple the specimens onto a gauze backing for easier handling, set them back into petri dishes, package the dishes in boxes, gas and seal the boxes to ensure germ-free air quality, and oversee the boxes' transport to Shriners. The technicians and postdocs in his lab rose to the challenge with a gusto he'll never forget, Green proudly relates. At the same time, there's every indication that Green himself proved an inspiring leader. Carolyn Compton, a pathologist now at McGill University who was part of the Shriners team that worked on the two brothers, recalls being struck by how "calmly and fearlessly" Green pressed forward in his endeavor to take a new concept straight to the patient. "At first, he really didn't know what he had," relates Compton. "But the minute he learned how to culture these cells and expand them in vitro, he played the iconoclast and was ready to try to apply them in the clinical setting. A couple of months with mice—and *boom*, right into patients."

In the early weeks, the children's red and raw bodies made for such a gruesome sight that Green sometimes wondered if he were doing the right thing. Wouldn't it be more merciful to let the boys die? But a nurse at Shriners said something that stopped his nagging doubts. Recounts Green, "She told me how one of the boys, just before receiving general anesthesia in the operating room, had pleaded with her not to let him die. As burned as they were, these children had a fierce desire to live, and so I became fully reconciled with trying to help them."

The Shriners staff would witness what amounted to a miracle. Small patches of skin salvaged from the armpits of each boy and expanded 10,000-fold in Green's lab meant the difference between life and death for each brother. In essence, the stem cells in those remaining underarm patches of epidermis saved their lives. Due to its resplendent regenerative capacity, this cell source would serve to replace fifty to sixty percent of each boy's total skin surface. Another source of stem cell would enter in as well. In skin areas that had escaped with second-degree as opposed to third-degree burns, stem cells that live deep in the hair follicle survived and were able to return those areas of skin to near normal.

Within a few weeks, both boys had achieved medical history simply by being alive. At first their thin grafts were delicate and vulnerable, but in the ensuing months the new epidermis became more securely anchored to the underlying connective tissue. It would require another two to five years, and yet the connective tissue beneath the restored epidermis would also heal and return to normal. "It was the first demonstration that human stem cells of any type could be expanded substantially in culture and used to permanently restore a patient's lost tissue," says Rheinwald. With a bone marrow transplant, it wasn't necessary to multiply the marrow's blood-forming stem cell, because the body saw to that. But in the case of widespread skin loss, growing up supplemental skin cells was necessary.

Having shown that real skin can be custom-grown in the lab, Green would soon be hailed as "the father of skin culture." Many in medicine would make note of this biologist who, working "far ahead of the curve," in the words of Pamela Robey, chief of the NIH's Craniofacial and Skeletal Diseases branch, was "developing techniques for tissue regeneration well before it was fashionable." Here, taking its place beside a bone marrow transplant, was another outstanding example of regenerative medicine that allowed for treating the body with its own cells. One distinguishing mark of Green's work was the way it so smartly overcame the main reason so few cell-based treatments yet existed, which Green today succinctly sums up as "a culture problem." Harvesting cells from the body, keeping them alive in culture, and expanding them constituted one of the tallest orders that regenerative-medicine scientists would face.

Green made his new skin-growing technique available to patients everywhere by forming the company BioSurface Technology, which in 1994 was purchased by the larger biotechnology firm Genzyme in Cambridge, Massachusetts. Genzyme's product Epicel continues to be sold and at present is the only commercial skin-replacement product of its kind for large-area burns. The harsh reality is that because so few people who are badly burned survive, Genzyme receives less than a hundred requests for the procedure each year, according to a Genzyme spokesman. With smaller burns, the kind most often needing treatment, the primary skin-replacement method continues to be skin grafting, according to Carolyn

Compton. "The skin graft is much too useful, too reliable, and too cheap to ever do without," she cites.

Green's method of growing basal-layer keratinocytes and crafting them into transplantable sheets of skin, while spectacularly innovative, has its weaknesses, Green himself admits. It is labor-intensive, given that it takes several weeks to provide a patient with skin. Plus the resulting sheet—all of two to three cells thick—is fragile and easily torn.

What comes as no surprise is that over the last several years many other researchers have tried to culture durable skin using both natural and synthetic materials. Surgeons would love not to have to rely on a patient's own tissue and instead have plenty of skin on hand for repairing skin. According to various authorities, some of the resulting formulas contain proliferating basal keratinocytes, which makes them therapeutically useful, whereas other formulas have proven inferior, their makers failing to base their concoctions on skin cells that have an adequate degree of stemness.

A major stumbling block, says Pamela Robey, is that there is still no easy way to select for epidermal stem cells, maintain them, and increase their numbers without some of them maturing and losing stemness. "And if you try repopulating a wound with engrafted material that doesn't contain stem cells, the effect is not going to be very enduring," she offers.

In the meantime, Green and Rheinwald's demonstration that keratinocytes of the cornea, esophagus, mouth, and genitals also could be significantly multiplied outside the body would spark the interest of other researchers, who are currently investigating a range of related medical applications. Here again, the secret of the culture lies in promoting the growth of only those keratinocytes imbued with stemness. Among the transplant procedures being worked on is a method for replacing mouth-situated epithelium that has been reduced by gum disease and mending tissue in the middle ear that has been lost to infection.

When last heard from, the two brothers whose lives were saved at Shriners Burns Institute three decades ago were still living close to their birthplace in the Midwest. One brother reportedly was doing fine, whereas the brother whose skin replacement had previously posed problems was said to have encountered related health issues. The long-term effectiveness of cultured epidermis is by no means a given, with many an associated complication unresolved. Howard Green, meanwhile, still marvels over the fact that a tumor in the testicle of a mouse put him on the road to culturing skin, whereupon he was able to offer this precious material to others.

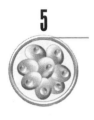

5

The Embryo Experiments

The concept of an embryo is a staggering one, and forming an embryo is the hardest thing you will ever do. To become an embryo, you had to build yourself from a single cell.
—Scott Gilbert, Developmental Biologist

By the early 1970s, the mood of biology was expectant—even heady, some might have said. The first gene ever identified in an organism had been written up in 1969—it belonged to the colon-dwelling bacteria *E. coli*—and it seemed only a matter of time before the genes and proteins of significant numbers of species would be decoded and biochemical explanations might exist for an organism's every twist and turn. Biologists appeared to be on the brink of understanding how Life worked.

But during these intoxicating times, which would extend well into the future, sometimes a more realistic note was sounded. As the French geneticist and Nobel laureate François Jacob cogently offered at a scientific meeting some years later, biology still operated at a relatively primitive, one-dimensional level. Although he and his colleagues were privy to the language of genes and proteins, there was so much more to learn about how these linear sequences "give rise to two-dimensional layers of cells, which fold in three dimensions according to a precise schedule, that is, according to a fourth dimension, time." Stressed Jacob, "We are three-dimensional objects, but we do not even have the concepts required to handle the behavior of such objects."

Jacob, who was based at the Pasteur Institute in Paris, made these sage remarks at a conference on teratomas and their stem cells that took place in the fall of 1982 at Cold Spring Harbor Laboratory. Nearly thirty years earlier, in 1953, at that bastion of biology on the north shore of Long Island, New York, James Watson had publicly announced his and Francis Crick's grand finding: the structure of DNA, the chemical that genes are made of. That was the year, as well, that Roy Stevens had encountered his first mouse tumor. In his talk, Jacob glanced back just twenty years, to when a meeting on teratomas "would have consisted of a dialogue between Roy Stevens and Barry Pierce." Stevens "literally invented" the system of using mouse teratomas and their embryonal carcinoma (EC) stem cells for exploring the parallel universe of the early embryo and its stem cells, Jacob recounted, while Barry Pierce "picked up" on this system and became the first to apply it toward "certain problems in embryonic development and cancer." Jacob saw these unsolved realms as the "two most important problems in biology," with a teratoma and its stem cells lying right "at the crossroads."

Jacob had led one of the first European groups that, in the late '60s, had rushed to investigate teratomas obtained from Roy Stevens's Mouse Strain 129. He and others had hoped that by studying an embryo and its changing cells through the substitute world of a teratoma and its stem cells, they might mount a grand inquisition into embryological growth, which had long posed such a mystery in mammals.

As odd a stand-in as testicular tumors from mice were, they beat collecting mammalian embryos. A mouse's early embryo (days 1-3) could be flushed from the oviduct, but it was difficult to collect enough of these specks for biochemical analysis. Then, once an embryo implanted in the uterus (day 4.5), it was hard to pick out, not lending itself to examination until some days later. Yet so much of interest was happening during these earliest days, when swiftly multiplying cells would have a profound influence on the final package. A mouse embryo's third day, in particular, was critical. Before the day was out, the blastocyst had formed, along with the cluster of stem cells inside it. The next day, as soon as the embryo implanted, many of these pluripotent cells started differentiating as punctually as Big Ben chimes, leaving their magic behind.

To think—these same master cells sat inside the bizarre little tumors of mice. What a heyday for developmental biologists!

In the late '60s, a box containing several of Roy Stevens's teratoma-bearing mice crossed the Atlantic, bound for a small laboratory connected to the venerable Department of Anatomy and Embryology at University College London. As soon as he heard about Stevens's unusual mice, Martin Evans had requested breeding stock as well as several transplantable teratomas from Stevens. Like other developmental biologists, he was full of anticipation at the prospect of having this alternative route into early mammalian development at his fingertips. Young mouse embryos were only "the size of a reasonably fat full stop," as he describes them today, and hard to collect from inside a mother mouse. When doing his doctoral study of frog genes at University College London, Evans had discovered that early frog embryos weren't much easier to work with. Their eggs might look commodious, but inside was mostly yolk. By comparison, Stevens's mice and their teratomas could be shipped straight to you, making for a much more convenient study.

By 1970, a handful of other scientists, including Gordon Sato and Boris Ephrussi, had shown that the stem cells—embryonal carcinoma cells—of mouse tumors could be kept alive in a dish, dividing and dividing. A standard culture medium was all that was called for. As much as Roy Stevens and Barry Pierce had pored over teratomas, they'd never grown their stem cells long-term; those they in-spected were mostly dead ones that were present in tissue slices they had shaved from the teratomas of dead mice and mounted on slides. And so, knowing that a teratoma's stem cells could survive in vitro was a very big deal, for only then would scientists obtain a realistic picture of how stem cells in mammal embryos grew and differenti-ated. To make EC cells flourish in culture, the researchers added feeder cells, just as Howard Green in Boston would do a few years later to ensure that his skin keratinocytes kept multiplying without differentiating. "Differentiation for a stem cell is the kiss of death; once it differentiates, it's no longer a stem cell," cites Evans. Knighted in January '04, Evans—a jaunty Englishman whose bushy eyebrows rival Francis Crick's—is now at Cardiff University where he directs the School of Biosciences.

Within earshot of the traffic along London's busy Gower Street, Evans plunged after key questions about EC cells, questions that Jacob later summed up at Cold Spring Harbor: "Do EC cells really differentiate in vitro? How many types [of cells] do they produce? Can they be induced to differentiate?" Science-related enigmas had attracted Evans since his toddler days. He wasn't much older than four when the sight of workmen at his family's home in Hertfordshire mixing water with cement had made him so curious and determined to find out why cement hardened that his parents had ordered him to keep his distance, for fear he would get in the way. He grew up loving chemistry and expected to become a chemist, until courses in high school and then at Cambridge drew him toward biology.

For the latest inquiry, Evans's jumping off place was Barry Pierce and Lewis Kleinsmith's crucial 1967 landmark experiment that dem-onstrated beyond a shadow of a doubt that the different cell types present in a living mouse's teratoma indeed arose from one cell and one cell only. The first order of business was to grow up large popu-lations of EC cells as a few other labs were also trying to do. Evans laced his EC cells with feeder cells and got them growing. Rigor-ously cloning, or reproducing, a single type of cell with the aim of getting a crowd of its progeny wasn't a snap, however. To begin with, cells are small and can glomp on to each other, and it was hard knowing you were starting with one particular kind of cell—in this case, a stem cell—which is what cloning a cell and getting pure populations is all about.

Arriving on the scene in 1973 was Gail Martin, Evans's first postdoctoral fellow, who had developed a knack for culturing cells while obtaining her Ph.D. in molecular biology from the University of California, Berkeley. Martin devised a method whereby she looked into a microscope and carefully picked up a single cell with a micropipette—making sure she had a stem cell, and not a stem cell stuck to a muscle cell, for instance—before dropping it into a well and growing its progeny in the company of feeder cells. Before long, the Evans lab had thriving batches of tumor stem cells that Evans and Martin kept undifferentiated by splitting the populations before they became too dense, just as Howard Green in Boston was doing with his skin cells. It was all part and parcel of learning how to manage cells outside of the body.

Next on the agenda: Would their flourishing EC cells differentiate into as many cell types in vitro as they did when inside a teratoma? Recollects Evans, "I plated out some cells, and I went away and left instructions with my tech to put more medium on the cells from time to time." More plasma, more vitamins. When he came back, a surprise was waiting for him. "The cells were differentiating *magnificently*—beating cardiac muscle cells, cartilage, nerve, skin cells, and more." As soon as the cells had started piling up, they began to differentiate, which was entirely consistent with what happens to stem cells inside the embryo when they start piling up.

Gail Martin reached the same destination by a different route. When she took EC cells off feeder cells, they aggregated into little round balls and similarly started to specialize. Roy Stevens had seen these clumps of cells, termed *embryoid bodies*, form when he injected stem cells into the abdominal cavity of mice. Staring into her microscope one day, Martin experienced one of those rare moments in a scientist's life when the fog completely lifts. "It suddenly clicked" that these clumps "were mimicking normal development," she recounts. Upon aggregating, they made an outer layer of endoderm, the same way that the early embryo's inner cell mass did after the embryo arrived in the uterus. That the tumor stem cells "had retained this knowledge" of what their normal counterpart cells did in the developing embryo "was mind-boggling," says Martin, who presently directs the Developmental Biology Program at the University of California, San Francisco. Here was more proof that a teratoma's stem cells and an embryo's stem cells were mirror images of one another, even though one type was normal and the other malignant.

By 1975, Martin Evans's London lab, as well as François Jacob's Paris lab, had demonstrated that a single cell in a dish could differentiate into an array of cell types. Knowledge was extending. "If aggregation stimulated stem cells to differentiate, preventing their aggregation kept them from differentiating," notes Gail Martin. Furthermore, the addition and withdrawal of feeder cells could significantly influence this outcome. Martin Evans would increasingly recognize the value of selecting the right serums and optimizing certain culture conditions, so that he could maintain control over his stem cells in culture. It was a balancing act of sorts.

Would the average pedestrian walking down Gower Street past the university's anatomy building have cared a whit that a few flights up a creaky wooden staircase scientists were able to grow undifferentiated cells that generated a potpourri of specialized cells? They might have cared had someone proposed the following. Given the fact that you can refurbish a lawn with grass seed or rebuild a forest with transplanted saplings, try imagining doctors repairing the human body with the very cells that compose it. Regenerative medicine was a fairly new concept, yet its antecedents were everywhere to be seen. One notable example was organ transplants, the numbers of which were on the rise now that researchers had begun to understand how to appease the immune system. The first successful human organ transplant had transpired at what is now the Brigham and Women's

Hospital in Boston as far back as 1954. A kidney had been transferred from one identical twin to another. As the 1970s lengthened, bone marrow and liver transplants were also on track and supported the idea that cellular tissues were transplantable.

The progress borne out in Martin Evans's London lab and at other benches created new challenges. Since researchers had stem cells from mouse tumors living in a dish, they now wanted to try their hand at directing these charmed cells. One wish, as seen, was simply to keep EC cells self-renewing without differentiating. As long as teratoma stem cells held to this pattern, they stayed pluripotent. Yet they seemed intent on differentiating, and indeed were genetically predisposed to do so. Another trick that researchers hoped to master was how to steer these stem cells down a specific path toward a certain fate. By the mid- to late '70s, they had identified various chemicals that could do this to a certain degree. Retinoic acid, for instance, could induce EC cells to differentiate into endoderm; dimethylsulfoxide changed them into muscle cells. This second trick, however, was a good deal harder than the first one.

Researchers were trying to make stem cells do things in a dish that they likely didn't do in their natural setting. Maintaining them in an undifferentiated state, for instance, "is asking them to freeze their differentiation program," notes Virginia Papaioannou. When put through hoops by researchers, stem cells in culture became artifacts of a contrived system, and couldn't be considered as normal as they once had been back in their real home.

Because embryonic stem cells from mouse tumors bore such a strong resemblance to those in embryos, for the time being scientists had plenty to occupy them. Still, the normal kind—stem cells that resided in mammalian embryos—clearly couldn't remain out of reach forever, not at the rate that other biologists were descending into the embryo. Some of these bold forays are worth a brief visit, especially because they put the spotlight on stem cells that much more brightly.

Better culture recipes and exceptionally fine needles and other tools invented in the late '50s and '60s enabled embryologists to test young mouse embryos in ways that, risky as they seemed, did not necessarily harm them. For instance, a minuscule eight-cell mouse embryo could be removed from its mother's oviduct, further grown and kept alive in a petri dish, and later returned to the uterus of a foster mother, who would give birth to the resulting baby mouse. The early mammalian embryo—a mere speck in a dish—was becoming accessible, which gave scientists ideas and the freedom to experiment. Now that egg cells, sperm cells, and embryos could be housed in dishes, experiments were inevitably headed toward in vitro fertilization, the ability to fuse egg and sperm in culture and create an embryo.

In the '50s and '60s as well, a longtime question about the embryo had found some answers. As an embryo grew and its embryonic cells lost potential—their fate pulled in the direction of one tissue or another—were their genes permanently lost or deactivated once they were no longer needed? Or did an individual's genes, an entire set of which exists in most every cell, remain useable throughout life, every gene capable of encoding a complementary protein at a moment's notice? If genes were lost, "it could explain how cells differentiated," notes Marie Di Berardino, an eminent developmental geneticist and cloning authority at Drexel University in Philadelphia. An early theory based on observations in worm embryos that "pieces might be breaking off the main part of a chromosome," with genes literally discarded each time a cell took another step toward a particular fate, imparts Di Berardino.

A useful test would be to extract the nucleus of a cell taken from an older embryo and put it into an oocyte, a one-cell unfertilized egg, whose own nucleus had been removed. Would the older substitute nucleus—really its genes—know to respond to chemicals within the egg cell that tell the nucleus to start dividing and catapult it into a developing embryo? If so, it would reveal that genes in the older cell hadn't been lost, but were still in place to set off all the bells and whistles that drive normal development.

Robert Briggs and Thomas King at the Institute for Cancer Research in Philadelphia (now the Fox Chase Cancer Center) put their shoulders to the experiment beginning in 1950. They took the nuclei of cells from slightly older frog embryos and implanted them into frog eggs. Indeed, something in the eggs' cytoplasm reprogrammed the nuclei and enabled them to start the growth of a new creature from scratch, and the historic outcome was perfectly healthy tadpoles. When Briggs and King repeated the experiment using the nuclei of cells from even older individuals, in most of these cases the development of new tadpoles ran amuck. Still, their progress left the impression that the cloning of creatures wasn't a ludicrous pursuit, at least as far as frogs and other less complicated animals went. Hardly anyone imagined mammals would ever be cloned. "In the beginning—in the '50s—there were all those jokes about cloning," recalls Di Berardino. "There were drawings of Tom King pictured as multiple people with a golf club in his hand. No, we never thought it would be done in mammals. The sole thing driving the work back then was basic research."

"I seem to remember that it was thought unlikely that mammalian nuclear transfer would be successful for a long time, if ever," recalls John Gurdon at the University of Cambridge. A mammal's egg was so much smaller than a frog's egg that replacing its nucleus would surely do it damage.

Yet it was Gurdon, when he was in the Department of Zoology at Oxford, who made the cloning of all kinds of animals that much more of an eventuality. In 1962, he reported that upon removing the nuclei of intestinal cells of tadpoles, which he took to be differentiated cells, and planting them in egg cells, many among them developed into normal tadpoles. There would be a long debate over this. Some colleagues suggested that perhaps Gurdon had inadvertently scooped stem cells out of the intestine, an organ rich in stem cells, and not differentiated cells as he thought. Not until forty years later—in 2002—would a research team vindicate Gurdon, proving that his previous science was without holes. *Differentiated cells truly did retain the use of every gene.* "Briggs and King, and then Gurdon showed us that a nucleus of a cell from a later stage was not irreversibly committed to what it was doing, but that it could go back and restart development. This gave us the concept that the nucleus retained the potential to do everything," provides Virginia Papaioannou at Columbia.

The modern era of cloning—or "nuclear transfer," as scientists refer to it—had arrived. The basic strategy is to move a nucleus from the cell of one individual into an egg cell taken from the same, or another, individual in order to produce a genetic replica of the cell's donor. Theoretically, the donor cell can come from any part of an embryo, fetus, or adult, although the older the donor cell, the less successfully its nucleus, when transferred, appears to be able to revert to a totipotent state.

Hans Driesch is often credited with being the first scientist to clone an animal, since he forced apart the cells of a sea urchin embryo and ended up with multiple individuals. This overlooks Abraham Trembley, however, who, when he cut a hydra in two and produced two hydra, was a bona fide cloner many years before Driesch himself was so much as an embryo. "No one had ever even guessed the possibility that a snip with scissors could make two animals from one," wrote Trembley's biographer John Baker. Briggs, King, Gurdon, and others brought a newer mode of cloning to the fore, however, one that introduced the idea of reproducing complex vertebrates from just one cell. This was a thunderous departure from the regular route of reproduction and its joining of two cells, one from male and one from female.

Still other experiments in the '60s displayed the extraordinary versatility of an embryo and its embryonic cells. Reminiscent of Driesch's 1890s demonstration that each cell of an early embryo has the pizzazz to become a full individual, in the early '60s Beatrice Mintz and Andrzej Tarkowski independently showed that you could combine two mouse embryos in a foster mouse's uterus and end up with one, and only one, normal-sized mouse, whose salt-and-pepper fur was the only visible token of its dual-embryo beginning. Animals made from two or more embryos were dubbed "chimeras," after the ancient Greeks' famous amalgam of lion, goat, and serpent. Like Driesch's flexible sea urchin embryos, Tarkowski and Mintz's chimeras were valuable proof that an "embryo's cells were not determined in what they could do; they responded to their environment," describes Virginia Papaioannou. Cells in the merged embryos had done whatever it took to end up as one normal-sized mouse, a sign of how tightly regulated an embryo in the uterus is and also how versatile its inner stem cells are.

Mouse chimeras made possible an experiment in the mid-'70s that completely riveted the attention of numerous biologists, gesturing as it did toward an ingenious new avenue into an animal's genes and the diseases locked therein. The experiment, already described in Chapter 2, was yet another example of the early embryo's flexible, alterable cells. Scientists injected stem cells from a mouse tumor into a mouse blastocyst, depositing them next to its inner mass of stem cells, and in a good many instances the added tumor stem cells, far from jeopardizing the embryo's growth, helped form a normal mouse pup. The tumor cells, instead of promoting tumors, fell under the spell of normal development. Here again was a case of cells, or really the genes scrunched inside them, changing in response to their environment. In other instances, however, the normal environment failed to matter; because of the added malignant cells, the embryo would acquire tumors.

While the public mightn't have been ready to appreciate this experiment, biologists saw it as a potential tunnel into a mammal's tens of thousands of genes, about which so little was known. Scientists might target a specific gene, modify it in a stem cell, put the stem cell into a mouse embryo, and thereby cause a change in the growing mouse. Any gene of the thousands present in each cell could be targeted, its function explored. Embryologists might be able to study genes that steered development; medical researchers might be able to knock out and study genes associated with a disease. Some distant day, maybe this same approach could be parlayed into fixing disease. "Even back then, the thought of using this technique for cell therapy was very much on our minds. That was a driving force," says Papaioannou, a member, along with Martin Evans, of one of the three teams whose 1974-75 work attempted this promising new approach. Although EC cells were malignant, it seemed safe to use them as vehicles for modifying genes, since when added to the embryo they appeared to contribute to the making of normal mice.

All of these marvelous visions depended on an EC cell's gene modification, once inside an embryo, becoming part of the animal's germ line—its eggs or sperm. That is, the modified stem cells would have to differentiate into, among other things, egg and sperm cells. Only then would future generations have this genetic change. Yet when attempting this goal, although one group claimed success, François Jacob, Martin Evans, and others had no luck. The egg and sperm cells derived from stem cells of teratomas couldn't start an embryo. Their chromosomes, scientists began to concede, must be intrinsically flawed—apart from any gene alterations inflicted on them. This realization caused researchers bitter disappointment, and work on this front would slow to a crawl by the late '70s.

Their probings of teratomas had proven so productive in so many ways, biologists had nearly forgotten that something might be wrong with the stem cells of these malignant growths. Now that they couldn't get past this fact, there was all the more incentive to do what no one had done: isolate the real thing, normal stem cells from normal mouse embryos. "When I moved to UCSF in 1975," Gail Martin recounts, "lots of people were trying to [fetch] stem cells directly from embryos. People were concerned that something was wrong with EC cells and thought it would be really great to make cultures of normal embryo cells." Scientists' newfound ability to grow a teratoma's stem cells was a big part of this readiness, according to Martin. If EC cells could survive in a dish, stem cells from embryos probably would as well.

The general location of stem cells in the early mouse embryo was no secret. But attempts in the late '70s by a number of investigators, including Martin Evans, who had relocated to Cambridge University's Department of Genetics in '78, and Gail Martin, who had been given her own lab at UCSF, kept ending in failure. Hampering their efforts was the scarcity of stem cells in an embryo's inner cell mass, as well as no clear idea of the ideal time to harvest these cells. Evans came to think that the right time was the far side of the blastocyst stage, "before or about the time of implantation," he mentions today, which added the hardship of hunting down and retrieving the dot of an embryo from the uterus. To "find this tiny patch of embryo" made for "a difficult dissection," he relates. A scientist also had to have prepared the proper culture to put the cells into once they were out of the embryo, or face losing them. Evans sought a culture mix that would encourage the stem cells to multiply, not differentiate; yet he was retrieving them at a stage when implantation signaled them to start differentiating.

In early 1980, Evans had a chat with a colleague at Cambridge that would prove to be a turning point. As he told Matthew Kaufman, a lecturer in anatomy and an up-and-coming authority on mouse development, he had been refining his methods, but try as he might, he still hadn't been able to sustain the growth of stem cells from mouse blastocysts in culture. The catch might be, he felt, that a blastocyst's inner cell mass simply provided too few cells to work with. If more existed, it might raise the odds of converting at least a few cells to long-running cell lines. "Martin was desperate to get cell lines going, and I offered him a suggestion—a trick," recollects Kaufman. There actually was a way to increase the number of stem cells in a mouse embryo's inner cell mass, he told Evans, a strategy that had been written up in a paper or two. It involved removing a female's ovaries immediately after she got pregnant. Without this source of estrogen and progesterone, the early embryo failed to implant and remained afloat in the mouse's uterus, its inner cell mass continuing to increase in size. It could wind up with double, triple, even quadruple the number of stem cells it normally has, according to Kaufman, who is now at Edinburgh University.

Kaufman's recommendation was right on the mark. By interrupting implantation in this fashion, Kaufman and his technician created blastocysts that in a microscope looked like Graf Zeppelins, as Kaufman describes them, or little sausages. These little oblongs were put into vials with tissue culture medium and carted the brief distance from Kaufman's lab in the Department of Anatomy across Cambridge's medical campus to Evans's lab in the Department of Genetics. There, Evans placed them in his own specially prepared culture medium, complete with feeder cells. A few days later, using a hair-thin bore pipette, he picked apart these tiny mouse embryos, dispersing the cells of their inner cavities in culture. After a few more days, there they were—little colonies of pluripotent cells growing away! He knew at once they were stem cells, he says, because "I was so used to looking at EC cells" in teratomas. By repeating the experiment, he eventually cultivated fifteen stem cell lines.

Evans's foresight, tenacity, and fastidious refinement of culture ingredients had paid off. For the very first time, science had embryo-derived stem cells from a mammal lying in a dish and "marking time," Evans observed some years later. When they were hidden away in the embryo, these cells were caught up in "an inevitable time progression," he described. "But taking them out into tissue culture, into the petri dish, they're marking time. And we can keep these cells, still with their ability to develop in any way, as millions and millions of cells in our petri dish."

An ocean and continent away, Gail Martin at UCSF was on track as well. Five months after Evans and Kaufman published their important news, a paper by Martin in the *Proceedings of the National Academy of Sciences* reviewed how she had reached the same milestone, albeit by a different route. Instead of beefing up a mouse embryo's inner cell mass, Martin had retrieved stem cells from the inner cell mass and jump-started them by growing them in a "conditioned" medium that teratoma stem cells had previously grown in, a technique she learned in graduate school. Proteins released by the teratoma cells made the embryo cells thrive. In her December '81 report, she referred to her cells as "embryonic stem cells," a name that would give way to the nickname "ES cells," paralleling a teratoma's EC stem cells, and one that would stick.

Neither of the special techniques that helped Martin Evans and Gail Martin with their pioneering work of isolating pluripotent stem cells from mouse embryos—implantation delay or conditioned medium—are necessary today. A scientist can now simply take an embryo out of a pregnant mouse, put its stem cells on feeder cells, and grow them. "Who knows why this didn't work to begin with," says Martin. Beginning experiments are always hard, she attests. And then one day, all the essentials fall into place.

The chief motivation for collecting normal stem cells from normal mouse embryos had been the hope that these cells could serve as vehicles for manipulating genes in embryos. Martin Evans persevered toward this goal, with long hours logged by postdoc Elizabeth Robertson and grad student Allan Bradley. (Robertson is now a principal research fellow at the Wellcome Trust Centre for Human Genetics in Oxford; Bradley, the director of the Wellcome Trust Sanger Institute in Cambridge, England.)

By the time of the Cold Spring Harbor conference in the fall of 1982, the one at which François Jacob made the point that biology was still at a primitive one-dimensional level, Evans and his lab crew were actually in the midst of finding their way down the rabbit hole into a second dimension. By '83, they had injected embryonic stem cells into a mouse blastocyst, whereupon the cells' progeny got into the mouse's germ line, and therefore into the mouse's descendants. By '87, they had extended this experiment. Altering an embryonic stem cell's gene in culture by means of a retrovirus, they then placed the changed cell into a mouse embryo. "So we had engineered the first mouse to carry a disrupted gene," cites Allan Bradley.

In the early to mid-'80s, two other scientists—Mario Capecchi at the University of Utah in Salt Lake City and Oliver Smithies at the University of North Carolina in Chapel Hill—had separately applied their wits to the feat of gene targeting, or replacing one form of a gene with a modified form. They had worked out an alternative way of installing a gene change, ingeniously hooking on to the fact that DNA is a twine of two corresponding threads, each gene having two copies. When introduced into a chromosome, a modified gene automatically seeks out its alter ego and lands in its proper place on the chromosome.

This long string of advances meant that biologists now had the supreme luxury of exploring the genome one gene at a time. The concept of a "knockout" mouse had been born. Virtually any known gene among thousands could be singled out, altered in an embryonic stem cell, and that stem cell could then be slipped into an early embryo. "We could now do real experimental genetics in the mouse," elaborates Evans. If you altered a gene, "What happened? What changed?" Today, more than 5,000 different knockout mouse strains exist that are helping investigators analyze genes connected to heart disease, inner-ear defects, anemia, and cystic fibrosis; genes that influence the risk for cancer; genes that make the immune system work; genes that regulate early development.

All this came about because back in 1953 a scientist on the

Maine coast bumped into a chinchilla mouse with an inherited burden. For scientists, there was something doubly engaging about that mouse. First, its teratomas got people interested in the wonderously versatile stem cells found in these tumors. Second, the early embryos of Strain 129 mice just happened to give rise to normal stem cells much better than embryos of other mouse strains, which led to these stem cells being used for the making of knockout mice. In fact, up until recently most knockouts were Mouse Strain 129 kin. If it weren't for these mice and the genomic depths they've opened up, observes Elizabeth Robertson at Cambridge, "We'd still be floundering around in the dark."

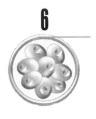

6

The Canary's Song

Stem cells can divide many times and don't get used up; that's what makes them stem cells.
—Fernando Nottebohm, Biologist

In the late 1970s, behind the towering wrought-iron gates that rise up just east of where 66th Street meets York Avenue in Manhattan, a birdman was closing in on a revelation that would signal a momentous "sea change," as one top-drawer scientist would call it. For so long, older neurologists had taught younger neurologists that the brain of adult vertebrates produced no new neurons. And then would come an out-of-the-blue finding, catching most people, but not everyone, by surprise. The amazement would soon lift, the disbelief blow away, and all that would remain would be a hungry incentive to try and steer yet another stem cell—this one from an organ thought to be nearly impossible to heal—toward therapies for countless grim disorders.

More than fifteen years had passed since Ernest McCulloch and James Till had snared evidence of a blood-making stem cell in adult mice. Following in their footsteps, investigators were at different stages of closing in on stem cells that populate other organs of grown animals. There were those in muscle that rebuild muscle fiber; those deep in the walls of the small and large intestine that replenish the intestinal lining; those in the basal layer of the skin that serve to replenish skin; and those in the testis that generate sperm cells. Sometimes an organ even contained more than one category of stem cell. For example, investigators were increasingly appreciating that alongside the stem cell in the bone marrow that made blood cells lived a cousin stem cell that spewed out new bone, cartilage, and fat cells.

Research on rats had indicated that the liver most likely had no resident stem cell. Of the two ways that organs acquire new cells— either from the division of stem cells or the division of mature cells— the liver's regenerative zeal seemed entirely due to the mature type. Liver transplants were demonstrating great promise by the mid- to late '70s, and when a person donated half of his or her liver to someone in dire need of this organ, both donor and recipient could rest assured that their portion would grow back to its original size, usually in about a month's time, thanks to the lightning-fast division of hepatocytes, the liver's highly specialized cell that processes nutrients from the blood for the entire body. If only every human organ were so regenerative! "That's why no one had been ready to believe there were stem cells in the liver—because hepatocytes were known to be the cells that responded to liver cell loss," relates Stewart Sell, a pathologist at the Ordway Research Institute in Albany, New York.

In 1975, however, Sell, who at the time was based at the University of California, San Diego, stumbled onto something. While investigating liver cancer in rats, he had run into a small round cell that was incredibly proliferative. It just so happened that during his med school days in Pittsburgh, Sell had had Barry Pierce as an instructor, and, "because of the seed that Barry Pierce had put in my head," he relates, he concluded that the rats' liver cancer probably was due to these small, fast-multiplying cells, which must be stem cells. "If it hadn't been for Barry Pierce, I would have gone along with the dogma," which had the cells originating in the bile ducts and proliferating in response to liver injury.

Later that year, Pierce, who had moved to the University of Colorado, happened to invite Sell to address a gathering of pathology students and faculty. Sell shared his interesting news for the very first time: He believed he had unearthed a stem cell in the liver. Given the liver's normally slow turnover of cells, he suspected that as stem cells go, this one was essentially inactive—certainly nothing like the intestine's busy stem cell. When part of the liver was surgically removed, mature hepatocytes, and not stem cells, reconstituted it. Nonetheless, he was quite certain that he had hold of a stem cell, one remarkably passed over all this time. Sell remembers that a hand shot up in the audience. "Why does the liver have stem cells if it doesn't use them?" a faculty member asked. Responded Sell, "Just because hepatocytes can regenerate the liver doesn't mean stem cells aren't needed." He and others would go on to identify situations in which they indeed were badly needed. Should the whole liver feel the severe effects of viral hepatitis or a toxic chemical, whereupon its hepatocytes dangerously stop dividing, then its stem cells would kick in, in an attempt to revitalize the liver with cell progeny.

By and large, adult mammals didn't stand out as regenerative creatures. Yet as the evidence grew that numerous organs had stem cells that spun out new cells, adult mammals were beginning to look more regenerative than anyone had ever stopped to notice. Search as they might, however, biologists found no signs of stem cells in the adult heart, pancreas, or kidney. And as for the brain, there was no point looking there for stem cells, at least not in higher vertebrates, most agreed. Ever since the great Spanish neuroscientist Santiago Ramón y Cajal brought the nervous system into fuller view in the early century, the conventional wisdom was that a mammal's neurons were made primarily before birth, during embryogenesis. Neurologists somberly advised, "Take care of the brain cells you are born with, because there's no replacing them." Or as Cajal had choicely put it: "Once the development was ended, the founts of growth and regeneration . . . dried up irrevocably."

People lived under the impression not only that their brain cells were irreplaceable, but that virtually thousands were dying every second. The brain's regression was thought to start in adulthood and accelerate in old age, the loss of neurons linked to the ravages of time and possibly made worse by such lifestyle habits as heavy drinking and drug taking, not to mention degenerative calamities like Alzheimer's or Parkinson's disease. The inescapable reality that disease and injury to the brain could be so deadly only reinforced the belief that this noble organ had no way of repairing itself. It seemed devoid of stem cells that might forge new neurons, and its mature neurons seemed incapable of division. How could they divide? scientists would exclaim to each other, given their often long and branching processes, the fibers (axons and dendrites) that connect them to any number of other neurons. What a tangle it would be if they did divide!

While it might lack stem cells, the adult brain was perceived to have a bit of regenerative oomph nevertheless. Neurons could sprout new connections and often did so when the brain was under duress. In addition, the nervous system had a second type of cell—glial cells—that divided throughout adulthood. Yet glia, which were looked upon as the "glue" that supports neurons, were ranked as fairly plebeian in comparison to the kingly neurons. Neurons were the conduits of everything learned and remembered, which strengthened the impression that they and their connections must be as fixed and irreplaceable as memories were fixed and lasting. "It was hard to imagine how brain cells and their processes could divide and give rise to other cells that would integrate into the existing circuit—especially in the hippocampus, which people thought was involved in memory. It would somehow screw up the acquisition of memories," observes neuroscientist Fred Gage. Or as Pasko Rakic, another neuroscientist, once observed, you would have new cells that never went to elementary school.

Not that researchers had never glimpsed *neurogenesis*—the ability to generate new neurons in the brain and spinal cord—in adult animals. They had, and for several decades. But either the animals reported on were "lower" on the totem pole, their biology deemed different from that of more complex animals, or the science behind these reports was viewed by other scientists as less than persuasive, which is what befell a string of reports in the 1960s by Joseph Altman.

Originally from Budapest and trained as a psychologist, Altman was working in MIT's Psychophysiological Laboratory when, challenged by the paucity of insights into the brain and behavior, he began examining the brains of rats. "Like everyone else," he recounts, "I believed that neurogenesis in mammals was a prenatal phenomenon"—something that largely happened before birth. Yet scanning the brain tissue of ten young adult rats, he glimpsed what he thought were new neurons. "The first paper I published was a question mark: *Are there new neurons?* In two to three years I was one-hundred-percent convinced there was neurogenesis after birth—even into adulthood in some regions." In 1968 he relocated to Purdue University, where he continued collecting leading evidence.

Altman was in a good position to go brain-cell sleuthing, early user that he was of a novel way of tracing newly made cells. The method worked off an injectable form of thymidine, a chemical that when introduced into the bloodstream was taken up by a cell's DNA as a cell prepared to divide into two cells. Rendering thymidine radioactive turned it into a tracer. Once inside about-to-divide cells, it ended up in the nuclei of the daughter cells, tagging these new cells and making them easily identifiable.

For more or less a decade, Altman ran one after another thymidine-labeling experiment on rats, cats, and a few guinea pigs, and when he checked his autoradiograms (film exposures that reveal the presence of radioactive thymidine in cells), time and again thymidine's peppery traces jumped out at him. As he communicated to the rest of the world through a series of papers, he was seeing new cells in the mammal cerebellum, a hind region involved in muscle movement; in the olfactory bulb, which aids smell; and also in the hippocampus, a region dear to memory. The new cells he was spotting weren't just glia, the brain's far-and-away most common type of cell. He also was seeing new *neurons*, the cells whose connections were regarded as the brain's force and fire. Of this he had no doubt, he says today, for he had full confidence in his tools and the incredible vistas of cells that they were showing him.

To other neuroscientists, some of Altman's sightings in the mammalian brain weren't outlandish. Studies had shown that neurogenesis could keep sputtering along for a while, postnatally. For example, cells in a human infant's cerebellum were known to keep dividing and making new cells for as long as two years after birth. Instead, it was Altman's claims of new neurons in places where they had seldom, if ever, been described in the brains of older mammals that flummoxed his colleagues. New neurons in the hippocampus and the cortex? These regions were the seats of higher intelligence, places where the circuitry was deemed far too complicated to allow for the production or incorporation of new neurons. Altman admits that his evidence of neurons in the cortex was inconclusive; but of those in the hippocampus—"I was without doubt."

So tantalizing were Altman's findings that other investigators scrambled to replicate them. No one came forward with completely convincing data, however. In the late '70s, Boston University's Michael Kaplan also published accounts of new cells in rat brains, but his tests were viewed as not being extensive enough. As for Altman, some biologists would outright dismiss the Purdue biologist's claims or view them as an exception to the ironclad rule that grown mammals experienced no new neurons. Conceivably, what Altman was seeing in rats was a phenomenon left over from leeches and other lower animals; or a phenomenon that lingered for a brief time after birth. The impression that emerged was, the phenomenon was nothing to dwell on, because it wasn't generally found in mammals. Altman himself feels that, in a field rife with competitors, his work was just plain ignored. "People acted like our papers, which we published one after another, had never appeared," he says today.

Altman and Shirley Bayer—Altman's wife and a first-rate neuro-biologist in her own right—would continue to study the process of neurogenesis in early development, making up their minds to do so with minimal contact with the professional community that had let them down. Data compiled by Bayer in the early '80s, which Bayer feels were passed over by colleagues, provided the strongest proof yet of the continuous birth of new neurons in the adult mammalian brain. She and her Purdue team reported in *Science* that in rats of ages one month to one year, neurons in the dentate gyrus layer of the hippocampus increased by roughly thirty-five to forty-three percent. Moreover, she also gleaned unmistakable clues that neurogenesis kept happening straight into a rat's old age.

Accounts of new neurons forming in the nervous systems of a bevy of nonmammalian creatures, including fish, amphibians, insects, and crustaceans, kept accumulating, as they had for some time. Scientists observed new neurons materializing in the respective brain centers of a cricket, a milkweed bug, a monarch butterfly, and a praying mantis; in the spinal cord of a stingray; in the retina of an old goldfish. Neurogenesis often appeared to be a lifelong occurrence in these creatures. Yet most scientists continued to assume that only less complex animals, mainly ones that were coldblooded and whose brains and eyes grew throughout adulthood, displayed neuronal growth.

The brains of adult mammals and other higher vertebrates make no new neurons. Fernando Nottebohm, the birdman behind Rockefeller University's tall front gates, had encountered this basic tenet while he was studying zoology at Berkeley in the early '60s. Now, in the late '70s, as he probed the brains of songbirds to learn more about the mechanisms that enabled them to sing, he had to heed this assumption. But the further his research progressed—both in a ground-floor lab in Gasser Hall and at the university's Field Research Center in Millbrook, New York—the more he questioned what he had been taught. It was dawning on him that the very bird strains he'd been attracted to ever since his boyhood roamings in Argentina might be telling him something quite different.

"I had a passion for birds as a child. I couldn't take my eyes off them," shares Nottebohm, whose face—whether it's his bushy eyebrows, direct gaze, or the shape of his mouth—contains something avian about it, although the rest of him—square-shouldered, with a distinguished deportment—is not at all bird-like. Nottebohm did most of his growing up in Buenos Aires, but his instincts as a naturalist were honed at his parents' second residence, a ranch situated on the fertile pampa of central Argentina that stretches south of Cordoba. His hobby there, he recounts, "was to go walking or riding through the country noticing birds and animals." Prolific numbers of South American songbirds, such as mockingbirds, Chingolo sparrows, and finches, beguiled the boy with their whistlings. Not to be outdone were guira cuckoos, burrowing owls, and scissor-tailed flycatchers that, while not songbirds, invariably caught young Nottebohm's eye with a certain chirp or curious habit.

During his sophomore year at Berkeley, Nottebohm had taken a course taught by Peter Marler, an authority on vocal learning in songbirds, and had come away spellbound. Vocal learning refers to a young bird's—or human's—ability to imitate sounds made by its elders. Not only are birds and humans both vocal learners par excellence, outdoing other animal groups in this regard, but a similar biology possibly underlies the ability in both to vocalize. Marler himself had been a student of the field's originator, William Thorpe, an animal behaviorist at the University of Cambridge. It was Thorpe, in particular, who had extended the use of the sound spectrograph, an instrument invented for monitoring sounds undersea during wartime, into a tool for rendering a visual display of birdsong on paper. The retooled instrument did away with the drudgery of transcribing birdsong by ear, says Marler, "and made it possible to grapple with all the detail, variability, and complexity of birdsong. I like to think it was the beginning of the whole science of birdsong."

So scintillating were Marler's lectures about how various animals had evolved specific traits in tune with their surroundings that he nearly hypnotized his students, Nottebohm relates. The Englishman's descriptions of why birds communicated and the sounds they used for this purpose were so "irresistibly interesting," says Nottebohm, that he made up his mind to devote his own career to songbirds. By looking into how vocal learning evolved in birds, he hoped to pocket new insights into the evolution of human vocalization, and maybe even gain fresh clues as to whether the acquisition of language had paved the way for human consciousness. (Despite this last hope, Nottebohm admits that after nearly forty years of studying vocal learning, he is none the wiser about consciousness. "We all believe it's there, but if you can't quantify a phenomenon, you have nothing to grasp," he remarks. "And with consciousness, there's nothing to grasp, because you can't weigh it or measure it.")

When Nottebohm was seven or eight, he somehow captured a white canary that had flown into his family's large garden in Buenos Aires. Placing it in a cage in the courtyard, he listened to his friend's song for years to come—and even ended up an imitator. When relaxed and wandering around Rockefeller's Millbrook field center, which he directs, Nottebohm easily breaks into a melodic whistling.

Now, in the late '70s, he often thought of his canary and how it had been a harbinger of things to come. More than the cheerful chaffinch, the slightly squeally zebra finch, or any other bird that he had so far studied, it was the effervescent canary that was providing Nottebohm with clues about how a bird sang. A few years earlier, in 1976, he had mapped out the canary's song system, which, located in the bird's forebrain, is comprised of two small yet distinct areas of neurons. Impulses conveyed by the large neurons in these regions travel down neural fibers and activate the muscles of the syrinx, the bird's vocal organ. Instead of involving cells throughout the brain, the biology of birdsong appeared to be quite confined, which was terrific news for Nottebohm and other like-minded researchers, since it meant that figuring out the biology behind birdsong might be that much easier.

A question that occupied Nottebohm, one that had been around for some time, was, why is it that male birds so often are the singers, whereas females sing much less or not at all? Of the world's approximately 4,000 species of songbirds, those that migrate are most likely to exhibit this male-female discrepancy. The behavior may have evolved because migrating males often fly ahead of the females and use song to mark their territory and attract a mate. Nottebohm wondered if some anatomical difference between singing males and nonsinging females possibly explained their vocal differences, and later in 1976 he and his first postdoctoral student, Arthur Arnold, confirmed as much. They observed that a male canary's song center was as much as three to four times larger than a female's. If a larger song center accounted for the male's singing ability, this size difference made perfect sense. Yet it also came as a surprise. Such a blatant structural difference between a female and male brain had never before been encountered in a vertebrate species.

Nottebohm wasn't entirely sure what to make of this finding, except that there had to be *more* of something in a male's song center—more blood vessels, or more cell connections, for instance—that made it physically bigger and supported the outflow of song. Could it be that there were more neurons?

From decades past came a handful of reports that he read with great absorption. The gist of these findings was that female canaries could sing as vibrantly as males if injected intramuscularly with testosterone. Bird importers in the '40s and '50s apparently had been wise to this trick and used it to hoodwink pet-store owners. Accounts tell of storekeepers complaining that they would sell a melodious canary to a customer, only to have the customer reappear and demand his or her money back, all because the canary had suddenly stopped singing. The storekeepers suspected the importers of boosting the female birds' testosterone levels, which made the females sing, but only for so long.

Since testosterone had this effect on female canaries, their brain's song center must be extremely hormone-sensitive, Nottebohm realized. But what exactly did the ruse of implanting testosterone do to the song center? Enlarge it, as the singing male's larger song system implied? But how? Did the hormone cause blood capillaries to expand? Did it make neurons physically bigger? Did it cause a significant increase in the overall numbers of glial cells or neurons? Nottebohm would quickly shake off this last idea. "People would have found it laughable," he says today, given the assumption that the adult brain made no new neurons.

To find out what was going on in the female's brain, in 1979 Nottebohm set about the logical experiment. He implanted capsules containing testosterone under the breast skin of several female Belgian Waserschlager canaries and waited for hormonal changes to ensue. Sure enough, just as demonstrated by other scientists, in about a week's time the lady canaries were singing as robustly as choir boys. Some weeks later he proceeded to euthanize the birds, remove their brains, and weigh and measure them. He couldn't quite believe his results; but then again he could. The testosterone that outwardly had turned the females into singers inwardly had more than doubled the size of the female's two forebrain song-center regions. It was the first time that a hormone had been observed to create such a gross anatomical change in the brain. Of far greater importance, it was eye-opening testimony that the vertebrate brain was far more flexible and mutable than people realized. Here again, the brain was making more of something. Could it be cells? If new cells were being born, that presented the incredible picture of a regenerative brain wherein dwelled proliferating cells that had stemness—neural stem cells. Nottebohm was beginning to warm up to the notion.

On the heels of that test, the Rockefeller scientist completed another experiment that would spectacularly sew together the gathering edges of his research. He was assisted by his wife, Marta, who often lent a hand with his experiments. In the field, for instance, she had sat confined for hours in a blind in Trinidad recording the bugling calls of orange-winged Amazon parrots. At Rockefeller's field center in Millbrook, which lay a few miles from the Nottebohms' residence and consisted of a large Tudor-style fieldstone gatehouse that stood between field and forest, chief among her talents was analyzing by eye the birdsong that a sound spectrograph imprints on endless yards of film, a meticulous job that is now done by computer. In addition, she edited her husband's writing into "understandable English," by his own claim. Overall, Marta was Nottebohm's tireless sounding board, "a refreshing, uncorrupted intelligence," as he describes her, whom he relied on more than he did the scientific community, which he viewed as being filled with theories that could lead one astray.

Ornithologists had long known that male songbirds sang most brightly in the spring, but as summer set in sang less and less, until their song all but disappeared in mid- to late summer, which was molting time. In the fall, a bird started getting up to song speed again and by January had acquired a whole new repertoire. A detailed comparison made by the Nottebohms of bird brains in spring and in fall brought the fascinating discovery that the canary's song center changed size according to the season: in singing season, namely spring, it was bigger; by early fall, when the singing season was over, it was smaller—much smaller. Over the course of those few months it could shrink by as much as half. The Nottebohms weren't terribly surprised to further discover that the birds' levels of testosterone rose and fell directly in parallel with the song center's swelling and shrinking.

When testosterone levels rose, what biological change swelled the song center, making a bird sing more? Testosterone itself wouldn't account for this swelling. Rather testosterone was causing something to grow and make more of itself. In various papers, including the above experiment's write-up in *Science* in 1981, appropriately titled "A Brain for All Seasons," Nottebohm aired his hunches. Maybe the song center's expansion was due to an increase in either the connections between neurons or the numbers of neurons or glia. His lab soon got evidence. In spring, at the peak of singing season, the brains of male canaries indeed had a greater complexity of fibrous connections than in late summer. Yet he suspected a proliferation of new brain cells was also happening. If the brain harbored new cells—or stem cells that gave rise to new cells—one might never know they were there unless the right markers were used.

Nottebohm's was "a lone voice" in those days, describes Steven Goldman, a former graduate student of the Argentinian. Few others were raising the possibility that the vertebrate brain might be fashioning new neurons. Moreover, Goldman adds, Nottebohm's familiarity with a bird's song system made him uniquely qualified to utter such an iconoclastic theory. Casting about for a laboratory to work in, Goldman, then a first-year student in Rockefeller and Cornell's joint M.D.-Ph.D. program, had been drawn to Nottebohm's lab both because Nottebohm's focus on bird brains was "so far out there," recalls Goldman, but also because of his own burgeoning interest in brains and the then-science-fictional prospect of repairing them. In his freshman year at the University of Pennsylvania, where he majored in psychology and biology, he had written a report on a Russian psychologist's attempts to transplant the heads of dogs. "I'll never forget the photographs of those poor two-headed dogs," says Goldman. "It really made me interested in brain regeneration." Nottebohm's bird research also took Goldman "back to West Philly," he says, where as a boy he would escape his neighborhood's roughneck street life by slipping away to the nearby confluence of parkland and golf courses to watch birds. "So on some subliminal level, Fernando's was an obvious attraction. It was an opportunity to look at brain plasticity and at the same time have an excuse to study birds," relates Goldman, who today heads the division of gene and cell therapy at the University of Rochester Medical Center.

Nottebohm's initial aim had been to get to the bottom of how birds learned to sing, but around the time Goldman joined his lab, the birdman was unexpectedly veering off into much bigger territory. When he surveyed the field center's many dozens of bright-yellow canaries tinged with green, which were housed in an outdoor aviary at the time, he was aware that he just might be looking at an animal whose brain—a brain that could easily be confused with a chickpea if you didn't have your glasses on, Nottebohm would tell students—had surprising things to say about all vertebrate brains.

With Nottebohm's guidance and encouragement, in early 1981 Goldman launched an experiment to solve the puzzle: Did a canary's song center get bigger because of added new brain cells? Or was this change entirely due to increased quantities of neuronal fibers? The plan was to implant testosterone capsules in female canaries; inject the birds with radioactive thymidine; wait a few weeks to give the testosterone time to do whatever it did to the bird's brain; and then inspect the birds' song centers. If doses of testosterone did cause an upsurge of new glial cells or new neurons—bingo!—the thymidine should make the new cells spottable. Since neurons supposedly didn't reproduce, Goldman's own hunch was that multiplying glial cells were the unknown element that was making the song center get bigger. Nottebohm took a more wait-and-see stance. He felt it was entirely possible that new neurons were behind the swelling, but so might be some entirely unforeseen phenomenon. Like a latter-day Abraham Trembley, Nottebohm didn't like pinning himself to any one theory.

The study was small—only four birds—and unfolded largely in Nottebohm's Manhattan lab. Three weeks out, Goldman euthanized the canaries, but it wasn't until several weeks after that—after flushing them with formaldehyde to preserve their organs, removing their brains, embedding the small orbs in paraffin, slicing them thinly, then mounting each slice onto a glass slide—that he was ready to begin the really interesting part.

Perched on a stool in Nottebohm's office one evening, Goldman adjusted the magnification on an Olympus light microscope from low to high, and with that small gesture visually entered a cross section of brain. His gaze immediately settled on a crowd of cells whose nuclei contained black speckles of radioactive thymidine. The cells included new glia, and—there was no doubt in his mind—new neurons. "I was sure they were neurons because they were large, large like the pyramidal neurons that populate the song center," says Goldman. Amazement kept him glued to the microscope until well past midnight, that winter night in 1981.

When Nottebohm saw the results, despite being ready for whatever outcome, he too was astounded, yet also vigilant. "I thought the profiles of the cells were very distinct and different from glia. There seemed to be no question we were looking at neurons," he recalls. "But I think I always left the escape hatch open, in case they were cells that were only neuron-like" and not the real McCoy.

This small study was a teaser. A larger study was sorely needed that could verify this remarkable sighting of new neurons and further address a slew of questions. If the new cells really were neurons, where were they coming from—the song center or some other region? The scientists had noted a sprinkling of new cells in the wall of the lateral ventricle, one of the brain's passages for cerebrospinal fluid. Were cells born there, and did they then migrate the short distance to the song center?

On June 10, 1981, Nottebohm and Goldman began essentially the same experiment over again, this time using eighteen birds. The study would become Goldman's Ph.D. thesis—as celebrated a thesis, as it turned out, as a young neurologist-in-training could wish for. By euthanizing testosterone-treated birds at varying intervals, some of the researchers' questions began to find answers. In birds euthanized early on, lots of new labeled cells were seen in the ventricle wall and no place else, whereas in birds euthanized a few weeks later, hardly any labeled cells were present in the wall, yet considerable numbers appeared in the song center. Quite possibly, new cells were born in the ventricle wall and then traveled to the song center.

This brain display was something! On any given day, the number of new neurons that a canary's brain was churning out was approximately one percent of the song center's neurons, which wasn't inconsequential. And while it had been difficult to imagine how neurons in more advanced brains could possibly divide to give rise to new cells without causing a tangle of connections, here in the bird was an answer. Immature cells—neural stem cells—that hadn't yet sprouted axons and dendrites were dividing, and their offspring were becoming neurons.

Another piece of this discovery was that the new cells seemed to originate in the ventricular wall. This same scenario occurs in the mammalian developing fetus when new neurons emerge from this wall and migrate in droves to their final work stations in the brain. But cell migration in the adult brain wasn't necessarily anticipated, because the brain wasn't thought to make new neurons. Then again, Joseph Altman had pointed to this very spectacle in the adult rat brain with some insistence, and in his papers mentioned others before him who had also observed cell proliferation in the adult brain, including *man's*. "These facts would suggest a high rate of cell proliferation in adult rats in [the subependymal layer of the lateral ventricle] and the migration of labeled cells to as yet undetermined brain regions," wrote Altman in 1965. In a few years, he determined that one place that cells migrated to in young-adult rat brains was the olfactory bulb, the brain's forwardmost point, and named this path the "rostral migratory stream"—rostral, as in *forward*.

The surprises didn't stop there for the Rockefeller scientists. New neurons even showed up in the song centers of females that hadn't received testosterone. This finding amazed the Rockefeller detectives the most, since it hinted that the vertebrate brain might be giving birth to cells all the time. Female canaries weren't altogether mute; sometimes they sang a little at the end of breeding season, when their testosterone levels were known to rise a bit; and they also issued a range of calls. Possibly the new neurons contributed to this vocal behavior, the scientists guessed.

If they were to challenge the dogma that held that adult vertebrate brains were incapable of regenerating neurons, Nottebohm and

Goldman knew they had better be dead certain that the new cells they were glimpsing in a canary's brain were really neurons. Searching back through the literature, they would unearth Joseph Altman's papers posting evidence of neurogenesis in rat, cat, and guinea pig. Initially they were disappointed; they had imagined they were the first ones to spy new neurons in adult advanced vertebrates. And yet Altman's research trek became helpful to them, and not only because his findings supported their own. Learning about the "phenomenal resistance" that Altman had encountered made them twice as cautious about how to proceed, remembers Nottebohm. "Neither Steve nor I wanted to make fools of ourselves. We didn't want to encounter what Altman had encountered."

As part of Goldman's thesis defense, two distinguished neuroscientists evaluated the song-center cells, and they, too, judged them to be neurons. Their opinions and Goldman's lengthy scrutiny with an electron microscope left the Rockefeller researchers confident that their finding was real. Writing up the particulars for publication, they were exuberant to think that their discovery might set the old dog of dogma on its ear. Even Nottebohm, a man of decorum, admits coming close to jumping up and down. "When we had what we considered strong data that new brain cells continue to be produced in adulthood, and in the widespread manner we saw in birds," he recounts, "I had no doubt that this was the nicest thing I'd ever do."

They announced their exceptional news at the Society for Neuroscience's annual meeting in the fall of '82. Their paper followed in April, by which time the skeptics were out in full force. As far as many authorities in the neurology world were concerned, Nottebohm and Goldman's news was too exceptional to be true.

But Nottebohm, in his genteel way, held firm. And when he spoke at conferences, he would pose the even more heretical notion that just maybe neurogenesis was taking place clear across the animal kingdom—even in the human brain. "I had little doubt that this would prove to be a general event," he notes today. "I didn't believe in a species inventing a trick like that and keeping it to itself." Yet the skeptics contended that the phenomenon that the Rockefeller team was seeing in birds must be restricted to lower species. Primates? Out of the question. They raised other concerns. How could new cells in the wall of the ventricle possibly know enough to crawl in the direction of the song center? How could the Rockefeller investigators be absolutely sure that thymidine, when injected, wasn't labeling glial cells as well as neurons, making it seem as though there were more neurons than there really were? Most important, did their supposed neurons walk the walk and talk the talk of neurons? Did they snugly integrate into the existing neural network?

A particularly strong gust of mistrust broadsided Nottebohm and Goldman in '85 in the form of a widely discussed paper in the prestigious journal *Science*. Its author, a highly respected and trusted Yale neurobiologist, was one of the two outside examiners of Goldman's thesis who had agreed that the canary brain appeared to be making new neurons. However, upon conducting a similar investigation of his own with rhesus monkey brains, the scientist reported that he had failed to find so much as a smidgen of evidence of new neurons in these primates. A fixed population of neurons was too crucial to the workings of learning and memory to imagine that neurogenesis occurred in adult advanced vertebrates, he concluded.

As time went by, Nottebohm and Goldman—Goldman having left Nottebohm's city lab in late 1983 to begin his residency at the Cornell-New York Hospital across the street—had reason to become even more certain of their evidence, at least in birds. In 1984, Nottebohm and his postdoc John Paton nailed proof that thymidine-tagged cells were honest-to-goodness neurons. The scientists literally impaled new cells in the brains of living canaries with ultra-thin electrodes and showed that they were as excitable as neurons ought to be and capable of relaying impulses. Nottebohm today feels that this was the experiment that "brushed aside all ambiguities. Suddenly people believed what we were saying!"

In case anyone didn't, in 1988 another of Nottebohm's postdocs, Arturo Alvarez-Buylla, made further progress into the adult bird brain. Astonishingly, neurogenesis wasn't restricted to the canary's song center. New neurons turned up throughout its forebrain, an area that included the memory-involved hippocampus and a region comparable to the mammalian cortex. Cells that likely were neural stem cells were dividing in the ventricular wall, and some of the resulting new cells were self-renewing, while others migrated to the forebrain and became neurons. This revelation wasn't unlike Leeuwenhoek's when he squinted through his hand lens and spied minuscule animals moving around in a pot of rainwater. "He gaped at their enormous littleness," writes Paul de Kruif. Alvarez-Buylla admits to gaping, too, the day that he and Nottebohm were staring into a double-viewing microscope and saw a parade of small, young cells migrating toward the forebrain. This cell dispersement showed up in the brain of every songbird and nonsongbird the scientists inspected. "There were no impediments in the adult brain stopping cells from migrating long distance," Alvarez-Buylla notes today. Was the same activity occurring in mammals? If it was, the implications for treating the human brain were tremendous.

Nottebohm would come to suspect that as new cells were added to the adult avian forebrain, older cells must be dying. Otherwise the brain would keep getting bigger and bigger. Scientists may have missed neurogenesis in adult vertebrates for this very reason; if the brain maintains a stable number of neurons, you'd have to be sitting inside a living brain with a movie camera to know that neurons were perpetually added and subtracted. "When I realized that we were dealing not only with adult neurogenesis, but with a system of constant neuronal replacement, that's the part that I found mind-blowing," relates Nottebohm. "We were used to having skin cells slough off and be replaced; we knew the liver could replace itself, and that the lining of the gut did the same thing. But here the brain was doing it!" Nottebohm theorized that while most of the songbird forebrain maintains a constant population of neurons, the song center grows larger or smaller due to seasonal variations that encourage or discourage its recruitment of new cells.

The Nottebohm lab would hunt high and low for proof that a bird's ability both to sing and relearn a complete repertoire each spring was a direct result of seasonally appearing new neurons. "Common sense almost demands it," notes Steve Goldman. Yet the only evidence so far that neurons can support a particular function is a correlation between increased numbers of neurons in a chickadee's hippocampus and the bird's seed-storing behavior.

Nottebohm would give a lot of thought to the basic tenets that needed reexamination because of his lab's revelation of new neurons in the bird. How memory works remains far from understood, although the theory that gets the most votes casts the synapses in the role of the chief repository for memories—a synapse being the juncture at which an impulse from one neuron jumps across a sliver of space into the extended arm of another neuron. Nottebohm was inclined to think that if neurons were being replaced, something tied to learning and memory conceivably warranted their replacement. Here he parted ways with Trembley and came up with a theory: Maybe neurons, and not synapses, are the centerpiece of long-term learning; perhaps neurons store memory in the form of changes in their genes, with new cells encoding new memories. "Our brain, I believe, has a limited capacity for acquiring new memories because of its size, and unless you can replace cells, you might come to the end of your learning potential," shares Nottebohm. "Neuronal replacement could represent the brain's attempt to remain young."

Very gradually Nottebohm's bird work would be accepted; Altman's rodent studies as well. "It took a long time for opinion to turn around, as long a time as it takes a ship to turn around," says Alvarez-Buylla. Of no surprise to Nottebohm, others would uncover evidence of new brain cells being born in many vertebrate brains— mammals, even. And the sightings of brain-cell regeneration and migration prompted researchers to realize there was now tremendous reason to try to check, or even reverse, a neurodegenerative disease. Fetal tissue transplantation, in which cells from the fetal brain are grafted into an adult brain that is under siege from, say,

Parkinson's disease, had attracteed research attention for years. But now the adult brain's regenerative nature offered a promising different approach. "There was a tremendous rush of excitement to think that you might be able to induce neurons to be produced in the adult brain, and that you might be able to direct them to diseased areas," describes Nottebohm. Could researchers ever achieve the trick of getting stem cells inside the brain to differentiate into the actual type of neuron that was therapeutically needed? It was as though the brain was a newly discovered planet open to exploration. Anything seemed possible.

It was the spring of '95. Fernando and Marta Nottebohm found themselves in Athens with a few free hours between planes and decided to pay a visit to the Parthenon. They were en route to Jerusalem, where Nottebohm had been invited to give a lecture at the Hebrew University covering all the special things that canaries had shown him about the vertebrate brain. By then, researchers were reporting evidence of new neurons in the brains of more than a few adult mammals—from tree shrews to cats to marmosets—although none yet had turned up in humans. Nottebohm would tell colleagues it was much more fun in the old days, when no one believed him.

Once atop the Acropolis, the couple sat down on a stone rampart to admire the city spread out below them. No sooner had they done so when a white canary—one exactly resembling Nottebohm's feathered friend from his boyhood—flew out of nowhere, landed beside the birdman, and began to sing. Remarks Nottebohm, who to this day remains awed by the encounter, "That was *really* scary!"

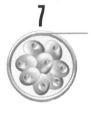

7

After the Birth of Louise Brown

*I guess I have never felt that knowledge is scary; it's
only scary when you don't use it for beneficial
things. I've never bought into the idea that knowing
something new is dangerous.*

—Ralph Brinster, Reproductive Physiologist

As the 1990s picked up speed, biologists continued to identify classes of stem cells that were hidden away in an
animal and to make better sense of a hierarchy that had been discernible for years. Stem cells in the early embryo gave way to others
in the developing fetus—among them, those in the developing testis
and ovary that perpetuated the species—which in turn gave way to
rare populations of stem cells that would renew certain organs and
tissues of the adult. The essential finding was that although stem
cells shared the ability to make more of themselves as well as generate specialized cells, stem cells were not all alike but differed in potential. Those in the embryo had immeasurable potential; those in
organs, by comparison, had lost something in translation.

The cell that sat atop this hierarchy in humans—the stem cell
inside the early embryo's inner cell mass—still lay frustratingly beyond the reach of biologists. This was the pluripotent cell that could
teach developmental biologists so much. It basically made the individual and every cell therein. One might have thought that because
Martin Evans and Gail Martin had blazed the trail by pulling this
cell from mouse embryos in the early '80s, its counterpart in hu-

mans would have been easy game. Researchers had pried similar cells from the embryos of hamster, mink, rabbit, pig, and cow. Yet isolating stem cells in human embryos required more finesse. Success would count on a healthy blastocyst with plenty of inner stem cells; precise timing in respect to the blastocyst's age; and a nutritive soup plus feeder cells that would keep the stem cells happily dividing in a dish while stifling their urge to differentiate.

One scientist who kept thinking about these coveted cells, and how they might serve as printing presses of sorts for a broad spectrum of mature cells, was Ariff Bongso, a fertility specialist at Singapore's National University Hospital. If anyone did, he had the instinct, reverence, knowledge, and resources for obtaining them.

Born in Ceylon (now Sri Lanka), the eldest son of a Christian Dutch mother and a Muslim Ceylonese father, Bongso started his career as a veterinarian specializing in animal reproduction, launched by a doctorate from Ontario Veterinary College. For a while, he investigated ways of identifying the sex of an early animal embryo. This skill, coupled with the ability to transfer an early embryo from animal to dish, where its genes were fussed over, and then back to the womb, could supply farmers with genetically superior milk-producing cows on the one hand and genetically superior beef-producing steers on the other. He also worked at improving the culture conditions that permitted an embryo to survive outside the body in the first place. As he became familiar with a wider and wider circle of embryos, ranging from cows to sheep to water buffalo, he marveled over "the notion that the common beginnings of gametes, fertilization, embryo cleavage, and growth" gave way to such an awesome diversity of species, he shares today. "The more I sank into embryology, the more it strengthened my belief that one and all are God's creations."

Soon Bongso made the leap into human reproduction, drawn by the great potential of in vitro fertilization [IVF], the bold new fertility treatment that bypassed low sperm counts, fallopian tube blockage, and other obstacles to pregnancy by fusing egg and sperm "in vitro"—literally, "in the glass" of a petri dish—in stark contrast to conception's normal setting, the fallopian tube.

The public had first been introduced to this daring procedure in the summer of 1978, two years after Bongso had gotten his doctorate. It was then that IVF's British pioneers, Robert Edwards and Patrick Steptoe, had announced the birth of Louise Brown, the world's first "test-tube" baby—more accurately, petri-dish baby. Angry protests had followed. In vitro fertilization's contrived merging of egg and sperm constituted an ungodly act, an unconscionable crime, some people charged. In the eyes of fertility expediters, however, the future had arrived. "I recall many accusations against Edwards and Steptoe, and how they were playing God," relates Bongso. "But I knew there was nothing wrong with this approach, because all the technique was doing was mixing a husband and wife's sperm and egg in a laboratory dish. I had faith these remarks would pass and the technique would become routine."

In Asia, Bongso was part of a small contingent that enabled IVF to become routine. In '83, he served on the fertility team responsible for Asia's first test-tube baby, and two years later joined the staff at Singapore's National University Hospital to help launch a state-of-the-art IVF program. "Singapore was the only Asian country to take the lead" in IVF, its university hospital having several visionary administrators, recalls Bongso. Before long, couples were singing the praises of fertilization in a dish, which gave them the children they otherwise might never have had, and interest in the hospital's new program "exploded," says Bongso. It wouldn't be long before IVF clinics the world over would similarly thrive.

For better or for worse, in vitro fertilization would radically change the public's perception toward conception and what an embryo could bear. Mouse embryos had been made to survive in culture since the '40s. This added twist of starting a human one in a dish, however, and then transferring it back into a mother's uterus seemed to some observers like one more step away from God and Nature.

IVF, however, caused no hideous consequences. Apparently the light of day did not deter a sperm from burrowing into an egg cell. Apparently a petri dish's plastic didn't do irreparable damage to a fledgling embryo. Louise Brown and other IVF babies weren't deformed, and the fear that when they became adults they might end up as characterless as the sterile dishes their embryos had started in proved groundless. Tens of thousands of test-tube babies later, most physicians today believe that conception in a petri dish poses no major health risk for the great majority of embryos. Two 2002 studies reported that a small percent of IVF babies may be more prone to either birth defects or low birth weight than those conceived naturally. Yet further validation of these claims is needed, with some fertility experts suggesting that if the findings do bear out, they might relate to the infertile nature of women and men who seek IVF and not the treatment itself.

Interest in the new IVF wing at Singapore's university hospital grew so rapidly that by the early '90s Ariff Bongso and four other specialists found themselves treating 800 patients annually. Meanwhile, Bongso, the unit's scientific director, was trying to improve the low survival rate of IVF embryos. When transferred back into the uterus for continued growth, only five to ten percent of embryos made it to birth. Fertility clinics everywhere were encountering the same low success rate, and to improve a woman's ability to get pregnant, a practice had sprung up. Fertility technicians had taken to creating, in vitro, a dozen or more embryos for each woman treated, whereupon usually three or more of these embryos were transferred into the womb at the same time. A woman then stood a fighting chance of having at least one embryo go the whole nine yards. It also meant that she had a greater-than-average probability of bearing multiple infants. (According to one report, an IVF mother has a twenty percent or greater chance of having more than one child.)

As for the embryos that were left over, they were frozen, which provided a potentially useful stockpile. Should the first transfer of embryos fail and no pregnancy ensue, a second and even a third

attempt could be made without having the woman undergo the un-
comfortable and expensive procedure of egg collection all over again.
Or, if pregnancy did occur, the unused, frozen embryos might translate into more children for the couple at a later date.

Stem cell research and the rising numbers of stored IVF embryos in the world would become so inexorably intertwined that it
pays to take a modern-day look at the system that has resulted in
millions of stockpiled human embryos worldwide. (Currently, the
numbers of IVF embryos that, upon transfer into a woman, make it
to a live birth are around twenty-five percent, although estimates
vary from clinic to clinic.) Douglas Powers, the scientific director at
Boston IVF, which bears the distinction of being the largest fertility
clinic in the United States due to its 3,000-plus IVF procedures each
year, provides this description:

"Suppose fifteen eggs are retrieved" from a woman, postulates
Powers, using realistic numbers. By exposing them to sperm from
the male, "we try to fertilize all of them, but usually only about
eighty percent—in this case, twelve—get fertilized." The eggs that
don't get fertilized get discarded. Of those twelve, continues Powers,
"often the three best embryos are transferred into the woman on day
3 of development." Healthier embryos are distinguished by evaluating each embryo's shape, texture, and other criteria through the microscope. With three embryos now in the uterus, that leaves nine
others. "Maybe two of those nine will have failed by then," says
Powers, and therefore are discarded. "Of the remaining seven, perhaps four will be good enough to freeze;" in which case, the remaining three will also be discarded. With so many embryos disposed of
along the way, IVF may seem like a wasteful procedure. Yet, according to fertility specialists, even in the natural setting of a woman's

reproductive tract, the first few days of human development are fraught with losses every bit as severe. An estimated two-thirds or more of egg-and-sperm unions are so abnormal that either they don't implant in the womb or, if they do, their development is short-lived.

At some future point many couples will use at least some of the embryos they've stockpiled. Many others won't. At Boston IVF, of the approximately 5,000 IVF embryos that are cryopreserved each year, roughly ten percent are left unused by couples, according to Powers. A survey published in 2003 of more than 430 fertility clinics in the United States counted nearly 400,000 IVF human embryos lying in a frozen state, a figure that will likely keep rising, since unlike in England, where a five-year storage clause limits the numbers of stored IVF embryos, in the United States surplus embryos are stored indefinitely.

Many scientists would come to believe that the mounting numbers of IVF embryos in freezers, of which a sizable number were doomed waste, would be a valuable source of stem cells. As some saw it, these beginning conceptuses were not even full-fledged embryos and yet they had the gift of Life inside them.

In 1991, Ariff Bongso boosted the sustainability of IVF embryos by growing them on a bed of fallopian tube cells, just as if they were sequestered in the fallopian tube, their home before descending into the uterus. Fertility specialists were generally grasping the fact that if they could mimic an embryo's natural environment in a petri dish, it would lead to healthier in vitro embryos. The particular approach that Bongso hit upon, he reports, gave IVF embryos a much better chance of reaching blastocyst stage on day 5, and, when transferred to the womb, of making it to a live birth.

In the process of growing fertilized eggs up to blastocysts,

Bongso's thoughts would instinctively turn to the inestimable power of the few dozen cells that lay in a heap inside the fluid-filled cavity of an early blastocyst. From these specks, each one being hundreds of times smaller than an ant's eye, entire civilizations had sprung. From these specks would sprout his clients' sons and daughters, whom he would run into years later on the streets of Singapore. Bongso maintains that his mixed Christian-Muslim descent made him acutely aware of the remarkable events of early human development, particularly the fertilized egg's power of procreation. "I grew up with the Bible and Koran side by side," he relates, "and reading the teachings of these two books in English, I found revelation science. In the Old Testament and the Koran it is described very well that the sperm clings on to the egg; and it's also stated that the embryo clings to the uterus. *Clings*—I was taken by that word," and left with a deep regard for the potential brought about by the interlocking of egg and sperm, embryo and womb. Now, years later, confronted with pluripotent stem cells inside blastocysts and so aware of all they represented, he was left to wonder—what can I do with these special cells to make Life better?

One day in early 1993, says Bongso, "It came to me in a flash." If he could isolate these unique human cells and sustain their division in culture, maybe he could control their fate. He would then have a gallery of cells with which to treat sick people. He raced to the library to do a search to see whether other groups had isolated human embryonic cells and also to collect clues about how to go inside the human blastocyst's balloon-like inner cell mass. Bongso's idea was "so powerful," the magazine *Asiaweek* later described, "that it seemed to come from God's own whisper."

Overseas and in the scientific literature, there were signs that people were slowly waking up to this idea of using stem cells collected from early embryos to regenerate the body's tissues. Having young embryos in dishes drew scientists' attention to an embryo's inner cell mass almost in a hypnotic way, making them ponder the pluripotent cells that sat there. For Bongso, because of the Seeds of

Life parables he was raised on and his study of animal reproduction ever since, "It was common sense that one should be able to direct stem cells into desirable lineages," he recounts: kidney cells for kidney disease; brain cells for neurologic disorders; pancreas cells for pancreatic ills. His hospital's active IVF program was a terrific resource. Thousands of embryos were stored away in liquid nitrogen tanks, many of them destined to languish in these cold beds and never be used. Bongso also had going for him the fact that Singapore's Ministry of Health tolerated human embryo research so long as researchers followed stringent guidelines. This policy was the envy of scientists in other countries, such as Australia and the United States, where research on human embryos was hampered by funding restrictions.

One of the Ministry's caveats was that if spare IVF embryos were to be used in research, they had to be less than fourteen days old. Up through day 13, it seemed safe to say, a human embryo has no capacity for feeling pain or any other sensation, for only on or around day 14 does the primitive streak start to form. Literally a narrow strip of migrating cells, this streak moves from one end of the embryo toward the other end like a creeping river, sending cells of the future skeleton and gut inside the embryo, and initiating the formation of the head and brain at the anterior end of the streak, explains Lewis Wolpert, a prominent British developmental biologist. Only at this stage, which is known as gastrulation, does the embryo begin to obtain the rudiments of a nervous system as well as begin to find shape. Some scientists believe that the primitive streak's onset represents such an important starting point in the development of human life that prior to day 14 the growing clump of cells can't be considered an embryo but instead should be looked upon as a "pre-embryo."

As a famous saying by Lewis Wolpert goes, "It is not birth, marriage, or death, but gastrulation that is the most important event in your life." Wolpert made this comment to "a doctor at a meeting in Belgium who clearly irritated me," he mentions in an email. He admits his remark may have been "a little exaggeration—but even so. . . ." While Wolpert extols gastrulation's importance in development, he's of the opinion that the embryo still has miles to go before becoming a person. "The embryo is not a person until it can survive almost unaided outside the mother," as he sees it.

After receiving approval for his project from National University Hospital's ethics committee, Bongso obtained twenty-one spare IVF embryos from nine patients who, having no further plans for them, were glad to donate them to the Sri Lankan's research. Bongso, meanwhile, was comfortable with the Singapore law he operated under; he, also, saw the primitive streak's onset on day 14 as the very first hint of a human being. As he grew IVF embryos up to day-5 blastocyst stage, then bathed them in an enzyme that dissolved their thin outer coat so that their inner stem cells would be let loose, he worked with the knowledge that these spare embryos probably would have been incinerated at some point in time. Why not, therefore, try to squeeze some benefit from them? Once he put the coatless balls in culture together with feeder cells, he essentially had cells from both the blastocyst's inner cell mass and outer cell layer in culture.

But all too soon came a big disappointment. The Singapore scientist had hoped that his culture broth would keep the cells multiplying, without differentiating into mature cells. To succeed for medicine's sake, he had to establish a stem cell line that would grow and grow. Explains Bongso, "You want those cells to produce millions and billions of cells, to increase their numbers for later differentiation into desirable tissues for treating patients." Yet quite maddeningly, although many of his cells lasted for two passages, they would spontaneously differentiate into mature fibroblasts. He and his labmates tried this, and they tried that, yet they couldn't hold the cells in check. Differentiation spelled the end of the line for a stem cell; it stopped being a stem cell and lost the magic to churn out specialized cell types.

To the best of their knowledge, nonetheless, Bongso and his crew had carved a new benchmark. Reporting their work in the journal

Human Reproduction later that year, they stated that this was the first time anyone had successfully isolated cells from the inner cell mass of "late preimplantation human embryos"—day 6 and day 7 human embryos that are about to implant in the uterus—and kept them alive in vitro for two divisions.

Had they really isolated bona fide embryonic stem cells, as Bongso today claims? Other researchers would say no; the cells probably were not stem cells, but more likely cells that had begun to differentiate and take a more specialized form. As other scientists were also finding out, it was relatively simple to separate cells from human blastocysts, but whether they were at peak pluripotency was another matter. Bongso ran a battery of tests on the cells but was unable to verify that they were supremely pluripotent, which is why he referred to them as "stem-cell-like" in his paper. Now he wishes that instead of rushing to publish his account, he had "patiently worked out the obstacles to crossing the second passage," he conveys.

Bongso, nevertheless—by keeping human embryonic stem cells dividing in a dish for any time at all—had managed to cross The Great Divide.

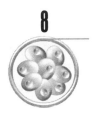

8

Monkeys to Humans

As he told me, he is interested in studying "how we start out as one little cell, a little teeny thing . . . and somehow that one cell turns into you."
—A second-grader, from his interview with James Thomson, Developmental Biologist

Stem cell scientists in the 1990s bore a certain resemblance to the hang gliders who soared back and forth over the hills of Madison, Wisconsin, in summer. Many would have liked to have been left in peace to swing on the wind and study this remarkable biological entity—a stem cell. But as the decade progressed, something like the force of gravity kept dragging them downward into an ethical, swarmy quagmire.

So it was with James Thomson, a young scientist at the University of Wisconsin in Madison. Thomas, in fact, had a hankering for aeronautic contraptions, including boomerangs, model aircraft, and gliders, and after arriving in Madison in 1991, he would have given his eyeteeth to have found time for hang gliding. Instead, he spent the better part of his days and evenings inside a lab at the university's primate research center bent over the cells of tiny monkey embryos that were only a few days old. A no-nonsense sort, he could be "relentlessly quiet," as one university acquaintance has described him. When he talked, his words tended to gallop, as if he wanted to get this interruption over with so that he could get back to prying out the secret of how one cell, then two, four, eight, and so on had the wherewithal to expand into trillions and fashion a monkey, or for that matter a person. The irony was that once Thomson's research took off later in the decade, this reticent scientist, who had no interest in the limelight, would find himself drenched in it, and also pulled downward by as contentious a debate as society had ever known—namely, at what point in early development should a human embryo be thought of, respected as, and treated like a person?

To be sure, it was an edgy time for the human embryo and fetus, and any associated research. The reasons are hard to separate, yet a few obvious ones stand out. In the United States, at the center of the issue was *Roe v. Wade*, the 1973 Supreme Court decision that made abortion a legal option for women, permitting the fetus's removal from the womb up to when it reached "viability." "'Viability' is somewhat a vague word," comments Norman Fost, a professor of pediatrics at the University of Wisconsin, Madison, "but what the court had in mind was the point at which the fetus can survive outside of the uterus, somewhere around six or seven months gestation." Noted the court in defense of its ruling, "In short, the unborn have never been recognized in the law as persons in the whole sense." Yet for those who believed the unborn embryo *was* a person in the whole sense, and obtained personhood upon conception or soon thereafter, *Roe v. Wade* was a flagrant offender of the commandment *Thou Shalt Not Kill*.

If ever a ruling drew a line in the sand! Groups that would increasingly be typecast by the press faced off: Anti-Abortionists, Right-To-Lifers, Devout Catholics, and Born-Again Christians crossed swords with Pro-Abortionists, Pro-Choicers, Liberal Thinkers, and Power Players for the medical industry.

Only after the *Roe v. Wade* ruling did legislatures begin "to systematically address the issue of research on conceptuses"—embryos and fetuses—Lori Andrews, a professor of law who specializes in biotechnology, has observed. (Just to reiterate, the first seven weeks of human development constitute the *embryo* phase; the last many months, the *fetus* phase.) The early human embryo's questionable status and whether lawmakers should allow researchers to utilize an embryo's cells made for a perfect political dog bone, bringing together as it did questions that forced science and religion to occupy a very tight corner. From the beginning, there was an unsolvable element about the debate, and over the years the centerpiece question wouldn't change. Is this morally and ethically aboveboard, or isn't it?

Further fanning the tense atmosphere that surrounded the unborn was the practice of transplanting tissue from fetuses. For decades, experimenters had regularly used the fetuses of amphibians, birds, rabbits, and rodents for their grafting experiments. "Slabs" of tissue were taken from specific regions of a fetus and transplanted to comparable regions in a recipient. The realization that fetal tissue was far more likely to survive uprooting than mature tissue raised the promising prospect of substituting diseased tissue with healthy tissue in human patients. Whether the tissue was taken from a fetal animal's brain, liver, or pancreas, it appeared to be graftable because it contained cells that, while they had committed to a certain organ, retained enough stemness to grow rapidly and generate specialized cells of that organ. An important bonus was that a fetal graft was less likely to upset a recipient's immune system than implanted mature tissue.

Scientists had experimented continuously with fetal tissue from small species, and much less with human fetal tissue. Interest in this practice was on the upswing in the United States, however. To the great delight of many researchers, a 1988 ban imposed by the Reagan administration prohibiting federal funds from fetal tissue transplantation research—research involving fetuses from induced abortions—had been lifted by President Clinton in January 1993, right after he took office. The very next month *Newsweek* spotlighted the "new hope" that fetal tissue held for incurable diseases, among them Parkinson's, Huntington's, and juvenile-onset diabetes.

Even during the previous period of restrictions, research into transplanting fetal tissue had carried on in the United States. Private funding had sustained it, along with federal monies available to investigators who worked with human fetal tissue from spontaneous abortions or stillbirths. Fetal tissue science kept chugging along overseas as well. Swedish researchers, for instance, had issued reports that raised optimism that cells removed from a specific area of the fetal brain might help Parkinson patients regain satisfactory levels of dopamine, the brain chemical that degraded in those patients.

Notwithstanding the bright outlook for fetal tissue grafting, people who were against abortion saw a moral landslide in progress. Each year in the early '90s as many as 1.5 million fetuses were being aborted in the United States; and with research that utilized fetal tissue gaining momentum, the fears unleashed by *Roe v. Wade* seemed more real than ever, that of fetuses being grown and aborted for financial or scientific gain, their value diminished to little more than farmed vegetables. To anti-abortionists and pro-lifers the practice of using aborted fetal tissue for medical measures seemed a duplicitous conceit—the killing of a life for the saving of a life. The sentiment would strengthen that new human life shouldn't be exploited. It seemed unjust enough that many thousands of IVF embryos were accumulating in freezers, and many more thousands of aborted fetuses were surgical waste. Raising further doubts about fetal tissue grafting were sporadic reports that it failed to help a disorder, or worse, led to tumors or a dangerous overgrowth of tissue in a recipient.

Lines were getting blurred and questions were piling up. In the opposite camp, Americans who supported human embryo research would increasingly question what they saw as a muddle of inconsistencies. If it was okay for fertility specialists to make so many IVF embryos that extra embryos sat in freezers for long periods of time and eventually were discarded, and if it was okay to test and study cells of aborted fetuses, then why wasn't it okay to put cells derived from leftover IVF embryos to better use? This is the ethical morass that James Thomson would be sucked down into later in the '90s.

While Ariff Bongso on the opposite side of the planet was isolating stem cell-like cells from human blastocysts in 1994 and multiplying them in captivity ever so briefly, Thomson, his supervisor John Hearn, his technician Jennifer Kalishman, and various others, all of whom were housed in Hearn's smartly managed Wisconsin Regional Primate Research Center, were in the midst of a stunningly long run. In hang-gliding terms, it was as though they'd been aloft for over a year. The previous year, they had managed to flush six-to-eight-day-old embryos from the reproductive tracts of pregnant rhesus and marmoset monkeys, whereupon Thomson had isolated stem cells from primate embryos and grown these peerless cells in the company of mouse feeder cells. The remarkable outcome was that the stem cells had been dividing and expanding in culture ever since, and *without differentiating*, thanks especially to the culture brew that Thomson had concocted.

This capture of stem cells from monkey embryos might well have happened later in time were it not for John Hearn, the British-Australian developmental biologist who, in a few short years, had turned the University of Wisconsin's primate center into what was widely considered the best in the world. "His goal of excellence was contagious," remembers Kalishman, a veterinarian who now works at Columbia University's Institute of Comparative Medicine. "He wanted to make it the most progressive primate center, and that included medical research, conservation studies, library resources," and the best possible scientists.

Ever since his graduate-school days at the Australian National University, Hearn's primary investigative niche had been examining how an embryo attaches to the uterus, or implants. By the time he was appointed to direct London's Institute of Zoology in 1980, he was a world authority on the subject, particularly in regard to primates and marsupials, although he investigated and marveled over the implantation process in a wide range of mammals. As he came to appreciate, the embryos of many mammals descend into the uterus between day 3 and day 5 post-fertilization, and yet when they actually implant varies greatly among species. The mouse embryo implants around day 4; humans and other primates, around day 7 or day 8; the wallaby, as late as day 20 (and then gives birth as soon as day 27). This small kangaroo and several other kangaroo species have a "fantastic" capability, shares Hearn. "The mother wallaby can keep a spare embryo in her uterus for up to a year, so if anything goes wrong and the young in her pouch is lost," the extra blastocyst "immediately activates," leading to a baby wallaby. "It's sort of like having a spare tire," describes Hearn, who now is back in Australia at the University of Sydney, where he is deputy vice chancellor. Researchers have slaved endlessly over the question of how to freeze embryos, he points out, "when here's an animal that can hold an eighty-cell blastocyst in suspension at body temperature for up to a year!"

Hearn first became aware of stem cells in his student days when he found out that they had something to do with why a salamander's lost tail "could regrow and shape as good as new," he relates. In the course of inspecting early primate embryos under the microscope in the early '80s, similar to what Ariff Bongso experienced, Hearn became conscious of the stem cells piled inside, cells that he knew were responsible for an embryo going from nearly nothing to something in a great hurry. "I would watch them hourly as the blastocyst settled down on the tray and its inner cell mass started to differentiate so quickly into the start of an eye, brain, muscle, and a beating heart, all within ten days. I was lost in the wonder of it," he recalls.

Martin Evans and Matthew Kaufman's extraordinary account in 1981 of freeing embryonic cells from mouse blastocysts made Hearn realize that his primate enterprise put him in an ideal position to similarly mine stem cells from monkey embryos. Among other things, with monkey embryonic stem cells in hand he and his crew would have the opportunity to engineer "knockout," or genetically altered, primates that could assist in better understanding genes and their role in the healthy or diseased body.

In London, Hearn began a focused program to retrieve stem cells in 1984, he recounts, and after crossing the Pond in 1990 and setting up shop at the University of Wisconsin's primate center, he began looking for the right person in this country to oversee the daily research demands of the stem cell investigation. A year later he found his man: thirty-three-year-old Jamie Thomson. "We clicked right away; we were right on the same track," remembers Hearn. "I could see that this guy was so flaming meticulous that he would persist and we'd go all the way."

For Thomson, who was keen on studying a primate embryo's earliest hours, a better opportunity couldn't have fallen from the sky. Raised outside of Chicago, by the time he was ensconced in the University of Illinois, Urbana-Champaign, this National Merit Scholar was already leaning heavily in the direction of biology as evidenced by his admission into the biology honors program. Two experiments that he read about in college had impressed upon him that an animal's early embryo was pretty astounding, because of its flexible stem cells. One was Beatrice Mintz's venture in the early '60s, in which she had combined the cells of two early embryos, and—behold—the mother mouse had given birth to a single mouse. The second was Ralph Brinster's '74 gamble in which his lab at the University of Pennsylvania's School of Veterinary Medicine took malignant stem cells from a mouse teratoma, placed them in an early mouse embryo, and—behold—the mother gave birth to a perfectly fine little mouse that had no sign of cancer. The inserted tumor cells somehow had been reset to normal.

Ralph Brinster, a reproductive physiologist with rapacious focus and drive, say colleagues, happened to direct a demanding joint-degree program at the University of Pennsylvania. Thomson applied and, top student that he was, was accepted. He would take away a doctorate in veterinary science in '85, a doctorate in molecular biology in '88, and shades of Brinster's intensity in the lab. There wasn't a finer scientist to emulate. Brinster's early culture techniques for embryos had been a pillar for the field. And then had come his savvy sense that maybe the embryo could incorporate introduced stem cells and thus be altered by external hands.

Thomson never entirely saw himself falling back on his first degree and becoming a vet. As he told an audience years later with a faint trace of humor, although he loved animals, he loved people less, and it was the people aspect of being a vet that he didn't think he could deal with on a day-to-day basis. More to the point, he says today, "the combined V.M.D./Ph.D. program at Penn is the best program of its kind in the world," and the joint degree was perfect for what he would become most interested in, which was researching human disease through the comparative guise of animal models.

The summer before Thomson began at Penn, Martin Evans's account of capturing stem cells from mouse blastocysts had been published. Thomson was attracted to the implications, as were so many young scientists. Here were normal pluripotent stem cells that might prove to be the vehicle for modifying genes in animals. (The idea of using these cells for therapies was not yet widespread.) Would Evans go a significant step further and isolate stem cells from human blastocysts? Or would someone else come along to claim that prize?

That same summer, the summer of '81, Thomson had been a Fellow at the Friedrich Miescher Institute in Basel, Switzerland, where, among other projects, he had tried to grow a whole plant from a single mature corn cell. This was the same as cloning a plant, or producing its genetic twin, which people had done for eons by taking a cutting of a plant or dividing one in two. Indicative of this long history, "clone" comes from the Greek word for twig, which is *klwn*. But cloning a plant by means of just one cell—and a *mature* cell, not an embryonic meristem cell—was a fairly recent accomplishment. Frederick Steward, a plant physiologist at Cornell, had been the first to achieve this feat in the late '50s, when he grew an entire carrot plant from a single mature cell taken from a carrot's thick orange root.

Similar to the attempts to clone frogs in the '50s, Steward's experiment hinged on genes inside a mature plant cell reprogramming and reverting to a totipotent state from whence could spring the different specialized cells and tissues necessary for the making of a new plant. Zoologists had begun speculating that the same biological process explained a salamander's ability to replace its limb: specialized cells in the stump *de*differentiated into embryonic cells, and the embryonic cells in turn *re*differentiated, generating all the necessary cell types for growing a new limb. Due to John Gurdon's cloning experiments in the '60s, it began to be recognized that the genome of at least some differentiated cells "is not irreversibly altered or blocked by differentiation," but could be reprogrammed backward to a younger potential, as one Michigan zoologist described in 1970.

While in Basel, Jamie Thomson never did get a corn cell to grow into an entire plant. Yet he appreciated more than ever the power inside a cell, whether a plant or animal cell. At Penn that fall, he began learning a great deal about the cells of a budding mouse embryo in Davor Solter's laboratory at the Wistar Institute. Solter and some of his crew were also studying Roy Stevens's mouse teratomas, and so Thomson took in a great deal about a teratoma's cancerous stem cells as well. Solter, a leading Yugoslavian-born mammalian embryologist, had done key experiments with these precocious cells when he was at the University of Zagreb. Planting a mouse embryo under the kidney membrane of a full-grown mouse—a place where researchers bury tissue to see what happens to it—he had checked back in a month and discovered that the embryo had lapsed into a teratoma, its stem cells differentiating chaotically.

And so at the bench Thomson was learning that there is a kind of similarity between stem cells in a teratoma and those in an embryo. "They can easily move one to another," as Solter notes today. Set in a strange environment, a mouse embryo's stem cells turned malignant; or, as in Ralph Brinster's '74 experiment, malignant stem cells turned normal when dropped into the surroundings of a normal embryo. Solter remembers Thomson as being a very diligent student, and he also remembers Thomson's proclivity for things aeronautic. Thomson was trying to master boomerangs around this time, and as Solter later noted, "I was never sure we would survive it."

Not until shortly after his graduation from Penn did Thomson learn how to harvest stem cells from mouse blastocysts. Through the suggestion of Solter, he did a brief stint with Colin Stewart, a British mouse embryologist then at the Roche Institute of Molecular Biology in New Jersey and one of the first scientists to replicate Martin Evans's commendable culling of embryonic stem cells from mice. Stewart guided Thomson through the work's trickiness. "Once you put a blastocyst in culture, you only had a window of time to pull apart the blastocyst and get stem cells," explains Stewart. "If you waited too long—which was a great temptation since as the embryo became larger one thought it was producing more stem cells—the stem cells were gone." They were that transient.

Thomson meanwhile had accepted a postdoc position at the Oregon Regional Primate Research Center in Beaverton, a top place to master primate developmental biology, and right before making his westward move, he and Stewart lunched together. "We chatted about how it would be interesting to [isolate] primate embryonic stem cells," Thomson recounts, "and he told me about efforts in Britain where they were trying for human ES cells but had failed up to that point, and how making them from primate was probably a good idea. That was actually the first time I'd thought about doing it in primate."

Thomson never went the distance of extracting stem cells from monkey embryos while he was in Beaverton. He was too busy with other projects—and spent the time, as well, hang gliding off high cliffs beside the Pacific. Moreover, he didn't have access to the numbers of quality monkey embryos needed for such an undertaking. His desire nonetheless grew to examine the starting hours and days of monkey embryos for the sake of having "a better model for human development," he mentions, which "convinced me that I really did need to derive primate ES cells as a more sustainable source of material." Thomson had heard John Hearn talk at meetings and was impressed by Hearn's first-rate primate outfit in Wisconsin, which featured a team that could routinely recover admirable amounts of high-quality monkey embryos. In the winter of '90-'91, while on

Christmas break and visiting his brother in Wisconsin, Thomson met with Hearn and was offered the job of pursuing stem cells. He began work that June, with the arrangement that he would also do his residency training in pathology.

Although Hearn's stem cell program had solid backing from Wisconsin's graduate-school dean and other higher-ups, a few of his campus colleagues took a dim view of it. Recalls Hearn, they felt that he was wasting funds and should stick with less risky projects that would bear fruit more rapidly.

Yet with Thomson aboard the project picked up speed. First on the docket was to flush adequate numbers of embryos from female monkeys without resorting to surgery. Thomson crafted a cannula— a long thin tube attached to a syringe, and the first ever devised for flushing embryos from the reproductive tract of marmosets—and it was up to lab tech Jennifer Kalishman to efficiently wield it. Once a hormone test indicated that a marmoset was pregnant, the monkey would be sedated, and Kalishman might dislodge anywhere from one to four minuscule embryos, which could only be spotted through a microscope. Extracting early embryos from rhesus monkeys was far more difficult, due to what Hearn describes as the rhesus's "tortuous" cervix and its "false passages." Luckily, Hearn's group included Steve Eisele, Wisconsin's talented chief animal caretaker who "had this incredible knack of almost feeling his way up the rhesus cervix" with a cannula, recounts Hearn.

On the receiving end of these tiny retrieved embryos, Thomson, with the aid of micropipettes and enzymes, would excavate out the stem cells and meticulously experiment with different culture ingredients to see what might keep these finicky cells multiplying, without differentiating, with an eye toward the methods that had enabled Martin Evans to maintain his mouse stem cells in a dish. Problem-solving was right up this Phi Beta Kappa's alley. "Jamie worked out all the details of the research on his own," recalls Kalishman. "He was extremely disciplined and not easily distracted, and had a beautiful, logical way of troubleshooting problems."

When Thomson wanted a break from work, the prankster came out to play. One particular prank still makes Kalishman laugh. "Jamie was always popping up and asking me, 'What are you doing? What are you doing?' Well, one day I sat down at my computer, and as soon as I struck the first key, this voice said 'What are you doing?' I looked around, but Jamie wasn't there. It was the computer! He'd programmed it to say, 'What are you doing?'" If Thomson was reticent in certain situations, it often had to do with his being his own person. Points out Kalishman, "He's not one to succumb to what other people think is popular." He wore the basics. He didn't own a television. And he did not suffer fools gladly.

As the months went by, Hearn gave his young racehorse more rein for managing the research's day-to-day progress, while he himself kept the project on track. "I raised funds, did the marketing, ensured quality, and ran interference against those who thought we were nuts and should give up," remembers Hearn. And that's how it came to pass that by 1994 the Wisconsinites not only had monkey ES cells in culture, but had watched them divide and survive for over a year. To test the cells' pluripotency, they had injected them into mice whose immune system was incapable of mounting an attack, and the cells gave rise to teratomas. Poking into these strange little tumors, the researchers spied "neurons, muscle, limb buds, hair follicles, tooth buds, and complex-layered gut," the whole jumble descended from stem cells. That made for "an eureka moment," remembers Hearn. They knew for sure they had uncovered the rhesus monkey's master cell.

On the eve of their rhesus report in August '95, Thomson sketched out to reporters why having monkey stem cells captive in a dish was such a big deal. Studying their differentiation would throw light on a monkey's development, he explained, and reveal parallel events in human embryos. Stem cells in human embryos were next in line to be isolated, and when they were, scientists would have the keys to transplantation medicine. Multiply them, coax them to differentiate, and you might have a constant source of mature cells with which to replace diseased cells. Yet in the end few reporters covered the Wisconsin team's rhesus paper. Stem cells in general were such a novel subject, and their connection to therapies for human disease a remote pie in the sky.

As the rhesus paper was headed for print, Thomson says, he and Hearn had no plans to go after embryonic stem cells in humans. "I imagined someone else would be there first," he recalls. Yet no such reports surfaced, and he began to think more seriously about such a venture. Hearn was supportive of the plan, but his own career was starting to take a new turn. In 1996, he and his family would relocate to Geneva, where he would join the World Health Organization's Reproductive Health Research Program as a senior scientist.

That left Thomson to decide whether to pursue the "mother" cell—human development's *sine qua non*. He had the know-how, having isolated the equivalent cell in monkeys. Another motivating factor was that the university had its own in vitro fertilization clinic. It wasn't exactly overflowing with spare embryos—Madison had a small population. And yet perhaps some couples would be willing to part with their unused embryos, especially for research that might liberate medicine from its heavy dependence on chemicals. Cell-based medicine was a concept that others could easily appreciate.

A missing cornerstone was funding. However, just as Thomson began looking for this necessary ingredient, Michael West, an entrepreneurial scientist, dropped out of the sky. "Things always seemed to work out for Jamie when he needed them to," says Kalishman. "I've always told my colleagues that he has a Midas touch."

Michael West was unusual in his combination of interests. He had studied the Bible as closely as he had the biology of a cell. He also had a flair for business and public relations, having made a small fortune building up and selling a family truck-leasing business in Michigan. Now, in 1995, he was in the process of turning a company that he had incorporated five years earlier into a headline-fetching West Coast biotechnology firm. Although his idea for the company initially struck some people as far-fetched, there are those today who call West a visionary because of it. He had come to believe that human aging wasn't necessarily an unstoppable train; there might be ways of braking its forward motion. A species' distinct line of immortality, its germ line, gave him this idea. Egg and sperm fuse to form a new individual, whose egg or sperm merges with another individual's sperm or egg, and so it had proceeded through the ages. Since its beginnings, the human germ line had never ceased; or, as West likes to point out, "You and I are made of cells that have no dead ancestors." Realizing this, he says today, "made me think we could outsmart aging if we could understand what allows for the germ line's immortality."

Geron—*old man* in Greek—was the name of his company, and its two-fold mission was to decipher the cellular basis of aging, no less, and to design drugs that stopped that mechanism. West, who had a Ph.D. in cell biology, had taken a leave of absence from medical school at the University of Texas in Dallas to start Geron in '90, but then had gotten so caught up in his company's progress that he had waved goodbye to medical school altogether. He saw the pieces of a puzzle starting to fit, he says. One piece was Leonard Hayflick's famous 1961 observation that most mature cells in the body divide only so many times before running out of steam. Another piece was research that linked a cell's telomeres—the ends of chromosomes—to the aging process: Each time a cell divided and its telomeres shortened, it took a fateful step toward terminating. The puzzle picture that leapt out at West was the telomere, and soon the main aim of Geron's scientists became trying to isolate telomerase, the protein that synthesizes telomeres. Inserted into a cell's nucleus, theoretically it might lengthen the telomeres—and a person's life. In reverse, if telomerase was inhibited, it might stop cancer cells from their ceaseless division.

But even as Geron's researchers were hard on the trail of telomerase, something else was occurring to West. During his second year of med school, he relates, "every Tuesday our pathology professor would bring in this stainless steel bucket filled with human tissue that was brought over from nearby Parkland Hospital," the Dallas hospital where, in 1963, doctors had tried to save the life of President John Fitzgerald Kennedy. The purpose of each bucketful, which might include anything from cancerous lung tissue to a liver damaged by cirrhosis, was to allow the students to see full-scale diseased tissues, which they otherwise saw only at the microscopic level on glass slides. One Tuesday, the professor appeared with something a bit different—a large teratoma removed from a woman's ovary. "We cut it open and in it was an incisor and molar among skin and hair and developing [cerebral] cortex," relates West. "At this point, I was interested in regenerative medicine, and my mind was burning with the question of, what cell does this? What cell makes all this? I didn't have a clue." Then he learned that a teratoma arose from a misguided stem cell, and that Roy Stevens at the Jackson Laboratory had shown that the misguided cell was a germ cell—or at least that was the case in mice.

Even once Geron, a telomere company, hit its stride, stem cells stayed on West's mind. So did an earlier observation by the biologist Howard Cooke that the telomeres of egg and sperm cells, or germ cells, didn't appear to shorten over time like the telomeres of other differentiated cells. Because egg and sperm were stem cells, West got this notion: If stem cells really were immortal and had long telomeres, he reasoned, why seek out telomerase itself to reverse aging in other cells? Why not instead try to harvest human embryonic stem cells with their long telomeres? If for the purposes of transplantation therapy you could coax ES cells to generate differentiated progeny, these mature cells would start off with long telomeres as well. You'd have *young* mature cells that could be used in the face of disease or aging.

West diverted two of his scientists to the task of helping him collect stem cells from sixteen-week-old fetuses that came from a San Francisco abortion clinic. By the mid-'90s, he relates, they had isolated stem cell-like cells but couldn't get them to grow, and it was at this point that West began checking around to see if other labs were making progress with human embryonic or fetal stem cells. Jamie Thomson's lab at the University of Wisconsin, Madison, was one of three labs he would learn about. The others were Roger Pedersen's at the University of California, San Francisco, and John Gearhart's at the Johns Hopkins University School of Medicine.

When Michael West, with hardly any advance notice, appeared in Thomson's Wisconsin lab in 1995, the timing couldn't have been better for Thomson. Having made up his mind to go after human embryonic stem cells, he had applied for funds "not through WARF"—the Wisconsin Alumni Research Foundation, the non-profit that provides licensing services for the University of Wisconsin—"but through what was called 'The Office of University-Industrial Relations,'" relates Thomson. "They turned down the ap-plication, and a week later Mike West visited." Thomson had heard through the grapevine that "they," the now-defunct office he had applied to, "were afraid of it," afraid, possibly, of the consequences of a Wisconsin researcher going into an area that had begun to heat up politically. West essentially made Thomson an offer he couldn't refuse. Geron would help fund the human stem cell work and mean-while license the University of Wisconsin's patent rights to its rhesus monkey ES cells. Human stem cells being steps away from the rhesus cells, Geron would do preliminary work with the monkey cells.

All that Thomson needed in order to proceed with his research on spare IVF embryos was approval from the university's twenty-four-member Health-Sciences Human Subjects Committee, the re-view board that stood watch over any campus research involving human subjects, born or unborn. He was fully conscious of where his lab work was carrying him—straight into the floundering, unre-solved debate in the United States over an embryo's societal status and legal entitlements. He, the hang glider, really was descending uncontrollably into a hazardous swamp. A wide range of people and organizations vehemently opposed the dismantling of early embryos. Coming from the opposite view, Thomson didn't view a human blas-tocyst as "equivalent to a human baby," although he had friends who did, was an employee of a large university that had its fair share of people who did, lived and worked in a state that had thousands of staunch Pro-Lifers who did, and was a citizen of a nation of mixed heritages, some of which did.

Even before his and Hearn's rhesus monkey paper was published, while still trying to decide whether to isolate human embryonic cells, Thomson had knocked on several doors looking for guidance as to what legal and ethical issues might arise were he to dismantle human blastocysts. One of those whose counsel he sought was Norman Fost, who, beside being a professor of pediatrics at the University of Wisconsin, also was the founder and director of the university's Program in Medical Ethics. "Jamie cared deeply about doing the right thing," recounts Fost. "He appreciated the larger social context and understood that others might be upset about what he was doing." It was clearly expedient for Thomson to approach Fost, who also directed the review board that either would or wouldn't allow Thomson's human embryo research to go forward. Yet Fost and other Wisconsin staffers maintain that Thomson went to far greater lengths to examine the moral ground he was descending into than was required.

It was a harsh coincidence that just when Thomson and his lab's monkey work put them in an excellent position to isolate human embryonic stem cells, their endeavor and its dependence on surplus IVF human embryos fell right "at the core of the right-to-life issues that had begun to permeate so many health-care issues," recounts Fost. William Jefferson Clinton, a Democrat, was in the White House, and in many respects more liberal days were dawning. At the same time, the powerful pro-life trend, which had been in progress since Ronald Reagan's presidency, notes Fost, "was dramatically intensified by Newt Gingrich's rise to power in '94." Right-to-lifers raised their voices in Congress and thundered from pulpits against the new biotechnologies that endangered societal mores and religious creeds.

From 1980 all the way to 1993, federal funding for IVF embryo research had not been available in the United States because of a defacto moratorium. Then, in late '94, the National Institutes of Health's Human Embryo Research Panel concluded that research utilizing excess IVF embryos appeared to constitute acceptable public policy, as long as such research was conducted on embryos before the primitive streak's onset on day 14. The panel's report was well received by an advisory committee to NIH director Harold Varmus, a visible proponent of stem cell medicine.

In '93, as seen, Clinton had cleared the way for federal funding for research involving fetal tissue. (Bear in mind that legislation tied to human *embryo* research progressed separately from legislation tied to research involving human *fetuses*.) And now the door stood ajar to create regulatory guidelines for researchers using human embryos. Yet it wasn't long before that opportunity slipped by. The following summer, that of '95, an amendment was introduced by two Republican congressmen, Jay Dickey of Arkansas and Roger Wicker of Mississippi, that summarily blocked federal funds for "research in which a human embryo or embryos are destroyed, discarded, or knowingly subjected to risk of injury or death." The Dickey-Wicker amendment, entered into the NIH appropriations bill, would take effect in 1996. According to Alta Charo, a law professor at the University of Wisconsin Law and Medical Schools and a member of the NIH's panel, President Clinton, due to the approaching midterm elections, had to make certain concessions and "couldn't take steps to the left."

Thomson's IVF embryo work would not be eligible for public funds due to the laws of the land. And yet such was the situation in the United States that his research could legitimately sail forth on the keel of private funds provided by Geron, a private company— that is, as long as Thomson got a thumbs-up from the university's health-sciences IRB, or institutional review board.

Norm Fost recounts that, upon receiving Thomson's protocol in 1995, Wisconsin's review board felt "a special burden" to diligently examine the pros and cons of IVF embryo research. If Thomson succeeded in extracting human stem cells, the research was bound to attract attention. Thomson's interest in staying with embryos that were less than fourteen days old made the board's job somewhat easier, day 14 being the earliest point at which pre-neural structures start to form and the embryo might experience anything resembling pain. In fact, Thomson didn't want to come anywhere close to day 14, says Fost. Nor did he have any desire to go the route of taking eggs retrieved from women, growing those up to blastocyst stage without fertilizing them, and harvesting stem cells that way. Although this had only been done with the egg cells of nonhumans, probably it could be done with human eggs as well, solely as an option for harvesting stem cells.

Wisconsin's review board spent most of its time on two questions, according to Fost. There was "the debate over the moral status of an embryo, but, secondly, whether the IRB needed to get engaged in that debate at all. We frankly concluded that we probably didn't need to, for several reasons." First, the board's members had differing opinions, and "endless discussion probably wasn't going to switch anyone's view." Also, notes Fost, "three distinguished national commissions had already looked at the same ethical issues in great detail"—the Warnock Committee in the UK (1984), the Canadian Royal Commission on New Reproductive Technologies in Canada (1993), and the NIH Human Embryo Research Panel in the U.S. (1994). "All had come to the same conclusions—that some kinds of research involving human embryos were ethically appropriate, within limits; and that whatever the moral status of the embryo, it wasn't comparable to a full human." The NIH panel had worded its decision thus: "Although the preimplantation human embryo"—the day 1 through day 7 embryo—"warrants serious moral consideration as a developing form of human life, it does not have the same moral status as infants and children." Among the reasons it cited for its decision were the early embryo's lack of nerve sensation and the very high rate of natural mortality at this stage.

In the years to come, the National Right to Life Committee, the Christian Coalition, the Coalition of Americans for Research Ethics, and scores of other groups across the country would unite behind the belief that an embryo was comparable to a full human being. The National Conference of Catholic Bishops, which had opposed the practice of in vitro fertilization since its beginnings, took tremendous issue with the idea that IVF embryos should be deemed a legitimate source of stem cells. To the Catholic Church, a human being comes into existence at the moment of conception and from that moment on deserves the respect and protection due a child or adult. Arguments that the embryo gains significance only after it descends into the uterus on day 7, or only after the onset of its primitive streak on day 14, or only after its heart begins to beat on day 22, or only after an accumulation of growth milestones confers an array of human traits were denounced as empty claims by Richard Doerflinger, the National Conference of Catholic Bishops' eloquent associate director of its Committee for Pro-Life Activities. Once a human life starts, Doerflinger and others in the Catholic Church contended, how can its beginning be any less important than what it grows into?

In a 2001 letter urging Congress to oppose using "live human embryos" for stem cell medicine, Reverend Joseph Fiorenza, the president of the National Conference of Catholic Bishops, wrote the following. "In his great novel *The Brothers Karamazov*, Dostoevsky raised the question whether it would be right to build a world without human suffering if 'it was essential and inevitable to torture to death one tiny creature' such as an innocent child to achieve that end. Each of us must answer that ultimate question in the depths of his or her own conscience. The claim that destructive embryo research will achieve such a utopian end is, we believe, a hollow promise. In the meantime, however, the killing will be quite real."

Yet others would not equate research involving an early human embryo with a "killing" or a "destruction," since in their eyes a cluster of cells doesn't a person make. Norman Fost holds the personal view, apart from any held by the university boards he serves on, that "a long list of reasons" exist for why a young embryo does not represent a person. Although its cells carry a unique set of genes that make the embryo human, "that doesn't establish it as a person," he maintains. Cells in your skin have the same genes, "but no one would say a skin cell or a blood cell is a person." An early embryo's cellular potential, he feels, doesn't make an embryo a person any more than an acorn's inner potential makes an acorn an oak tree. Conversely, just because a person has the potential to die doesn't mean they are dead. "An embryo doesn't look like a person," remarks Fost. "Nor does it experience suffering, which is part of why we care about what a person is. And even the most ardent advocates of an embryo-as-person don't ask that embryos be counted in the census; they don't ask that they be included in the tax code as deductions; they don't ask that they be covered by health insurance, and they don't ask that they have funerals."

At the end of the day—it was July 1995—after reviewing the decisions reached by the United Kingdom, Canada, and United States commissions in close detail, the University of Wisconsin IRB realized that "it wasn't likely to reach a different conclusion," relates Fost, and James Thomson got word that he was free to pursue stem cells in IVF embryos. Once the particulars of the Wisconsin-Geron collaboration were worked out, Thomson sent Geron a sample of the embryonic stem cells that the Wisconsin crew had procured from rhesus monkeys. Geron's scientists analyzed the cells' telomerase, the protein that keeps telomeres long and healthy, "and saw that they had more telomerase activity per cell than any immortal cell ever studied—more than even a cancer cell," recounts Michael West. "We were looking at the first naturally immortal primate cell."

The immortal master cells of the human embryo were all the more anticipated.

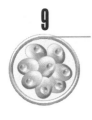

9

Epic upon Epic

The field of developmental biology has changed dramatically. The problems are the same, the questions are the same, but the tools available to us in the molecular age are of a different order of magnitude.

—Virginia Papaioannou, Developmental Biologist

The days of the unknowable human embryo were swiftly departing. Now that scientists could merge egg and sperm in the shallow of a dish, and quite literally *make* human embryos, the earliest hours of human development were open to unprecedented study. And meanwhile other novel techniques and tests, such as amniocentesis and genetic analysis, allowed conceptuses to be scrutinized in the womb as they grew to fetal stage and beyond.

This all goes to say that by the time Jamie Thomson decided to strike out after stem cells in human embryos, these enormous little cells were looming a whole lot larger in scientists' minds than ever before, and the small circle of teams striving for them—in Singapore, the United States, the United Kingdom, Europe, and Australia—was expanding. On China's southern coast, an immunologist at Sun Yat-Sen University in the city of Guangzhou was also ready to take the plunge. Professor Shunong Li knew this master cell had plenty to say about how the human embryo grows. He had faith as well that it might cure Mediterranean anemia, an inherited blood disease that seemed to show up time and again in children in south China, stunting their growth and sometimes seriously injuring their organs. Transplanted umbilical cord blood, or really the stem cells in cord blood, could alleviate this anemia. Yet sufficient quantities were hard to come by, which is why, Li believed, stem cells from embryos could prove priceless. Inducing them to differentiate into large quantities of blood-specific cells might be the answer to supplying these stricken children with all the cells necessary to make them well.

In England, sporadic attempts to extract stem cells from human embryos had been going on since the 1980s. The handful of investigators who had given it a try were mainly interested in knowing what stem cells could show them about early development. Christopher Graham at Oxford, for instance, set out in 1987 to investigate factors in human embryos that regulate the growth of their potent cells. "On paper, probably nobody was trying to isolate these cells for therapeutic purposes," notes Graham in an email. However, "all of us would have been exceptionally stupid not to realise their therapeutic value." In-vitro-fertilization pioneer Robert Edwards relates that as a young geneticist based at Glasgow University in 1962, he'd been enthralled by stem cells in rabbit blastocysts and their willy-nilly production of cells of all adult rabbit tissues. When he and Patrick Steptoe later teamed up to create human embryos in petri dishes, "their inner cell masses . . . screamed stem cells!" recalls Edwards. He envisioned a day when stem cells might repair human organs, but in the end decided to discontinue his group's stem cell work in order "to save all [IVF] embryos for their parents."

It was tempting to wonder how far along biologists could possibly be in understanding human development, if by the mid-1990s the talented cell that was so central to an embryo's maturation remained at large. Investigators blamed their failure in this regard on the poor quality of the IVF embryos they worked over. The very best ones were transferred into a woman, while the remainders placed in cold storage could be inferior. For all anyone knew, long-term freezing possibly diminished the quality of IVF embryos still further.

The knowledge that other researchers were having such a tough time isolating stem cells from IVF embryos had helped convince John Gearhart at the Johns Hopkins School of Medicine in Baltimore, who by 1995 was in the midst of formulating his own plan to try and acquire human embryonic stem cells, that it wouldn't be such a bad idea to visit an entirely different source: the fetus as opposed to the embryo. Whereas James Thomson was gearing up to go after a stem cell that sat in the hollow of a roughly six-day-old embryo, a leftover product of in vitro fertilization, Gearhart was aiming to isolate primordial germ cells—precursors to egg or sperm—from the gonadal region of an aborted eight-to-twelve-week-old fetus. The first: a dot of an embryo of not more than a few hundred cells that, while previously frozen, bore the potential to be returned to the womb and grown to birth. The second: a young fetus whose attainment of life had previously been ended.

Those differences aside, Gearhart, the director of research for the school's obstetrics-gynecology department, was pretty sure that an early germ cell from the fetus should be every bit as pluripotent and capable of producing an array of cell types as a stem cell from an embryo. His research would add an interesting twist to stem cell biology. Scientists, not to mention policymakers and church leaders, would increasingly agitate over the pros and cons of deriving stem cells from spare IVF embryos as opposed to the organs of grown people, and here, off to one side, was Gearhart hovering around a cell from the fetal gonads. Its pluripotency, although requiring proof, was entirely plausible since this cell type became the egg and sperm that perpetuated the human race. Moreover, this was the same cell that Roy Stevens, an older friend of Gearhart's, had shown could take a wrong turn in mouse embryos and grow into a testicular teratoma swollen with differentiated cells. It certainly proved its pluripotency in that instance.

Gearhart's younger years leave the picture of a shy, inward boy with a shock of strawberry-blond hair, who, after living behind the high stone wall of a Philadelphia orphanage from age six to sixteen, went to college hoping to become a fruit-tree grower and ended up with his master's in plant genetics from the University of New Hampshire; then earned a doctorate in fruit-fly genetics from Cornell; then became immersed in studying Roy Stevens's famous mice and their precocious stem cells; and finally came home to his own species—human development. Along the way, time after time Gearhart had the good fortune to cross paths directly or indirectly with distinguished sorts whose discoveries greatly added to the fast-moving tide of the biological sciences. Embryology, for one, was changing from a sleepy science full of fate maps—maps that trace cells from their origin in the embryo down a specific tissue pathway—into a more experimentally active and deductive discipline. The very term *embryology* had been growing stale since the '50s, gradually giving way to *developmental biology*, a phrase that recognized that a creature's development extended well beyond the embryo stage and would require the muscle of many disciplines if all was to be revealed.

When Gearhart went to Cornell for his doctorate, working a few floors above him in the Plant Science Building was Frederick Steward, the botany professor and larger-than-life Cecil B. De Mille figure. "We only do epics," Steward told his technicians about the experiments they undertook. Steward by then was famous for his masterful epic a decade earlier, the growth of an entire carrot plant from a single mature carrot cell. He had used milk from coconuts for his culture brew. Another scientist had discovered that orchid cells multiplied beautifully in coconut milk, and Steward "was trying to find out what factor in the milk promotes the development of cells," relates Marc Cathey, president emeritus of the American Horticulture Society and a student of Steward when the carrot drama was unfolding. Coconut milk being the impetus for the experiment—and not the cell or its inward potential—Steward struck it lucky. At the time, "The city of Miami was having terrible trouble with coconuts falling on tourists," recalls Cathey, and Steward, seeing a good resource, "had them collected and sent to him."

Coconut milk's nurturing factor would eventually be identified. But meanwhile, Steward's precedent of cloning a carrot plant from just one of its cells, and a specialized cell to boot, drew the research world's attention to something already known but sorely under-appreciated: Totipotency was not restricted to cells in the early embryo. Mature cells from adult plants could be cajoled to *de*differentiate into totipotent cells! They could be reconfigured to a blank slate.

A cell's versatility in this regard harkened back to the cell theory that Schleiden and Schwann had postulated over a century earlier. What force accounted for an organism's growth? Maybe it was central to the whole organism, as fat is to butter, Schwann had supposed. A second possibility was that each cell in an organism "possesses a power of its own, an independent life, by means of which it would be enabled to develop itself independently." In lower plants the latter seemed to be true, since "any given cell may be separated from the plant, and then grown alone," Schwann reasoned. He came to the conclusion that every plant cell probably has an "independent vitality," an observation that has been verified by contemporary plant scientists: Every single cell in a plant has the means to grow into an entire plant.

Although he never studied with Steward, Gearhart thought this was the greatest thing going in biology—to take a mature cell, render it totipotent by putting it in the right culture, and use it to yield a whole new plant. Finishing Cornell, Gearhart set his sights on doing his postdoctoral training in the lab of Jean Paul Nitsch, a French plant physiologist who had performed yet another epic when he had grown a tobacco plant from a grain of pollen. The equivalent in animals would be starting an embryo with only a sperm cell. But a few months before Gearhart was to leave for France, Nitsch was tragically killed in a car accident, and, in the fall of '70, Gearhart ended up instead in the laboratory of Beatrice Mintz at the Institute for Cancer Research in Philadelphia, where, for the next five years, he would probe the weird tumors of Roy Stevens's famous 129 mouse strain and their precocious stem cells.

Once again, Gearhart found himself working in a building steeped in cloning history. Less than fifty feet from his bench, Robert Briggs and Thomas King had set the stage for all future animal cloning with their 1950s experiments. Using cells from leopard frogs, they replaced the nucleus of an egg cell with a nucleus from a frog blastocyst. Fooled into thinking it had been fertilized by sperm, the egg cell launched into division and developed into a tadpole. Similar to Steward's carrot drama, a donor cell's transplanted genes were persuaded into a totipotent state, from whence sprang a genetic replica of the donor. With Steward, it was the coconut milk that had inspired the cell's genes to *de*differentiate, or roll backwards to totipotency; in the case of Briggs and King, it was the egg cell's cytoplasm. It was as though someone took a bath, and something special in the bathwater returned them to infancy.

Beatrice Mintz, in whose laboratory Gearhart was doing his training, was carving major benchmarks as well. Years back, she and Tibby Russell at the Jackson Laboratory had determined the migratory route of primordial germ cells in new mice embryos, a finding that had helped Roy Stevens prove that errant germ cells started teratomas growing in mouse embryos. Her reputation had soared in the early 1960s when her team had "glued" together two mouse embryos in a dish by removing their outer membranes and letting their cells grow together at 37°C. When she put the resulting single embryo into a surrogate mother, the mother had born a pup whose two-color striped fur signaled its two-embryo beginning. (A Warsaw University team headed by Andrzej Tarkowski published similar results.) Valued as one of biology's all-time classic experiments, this caper had fed the imagination of Gearhart, Thomson, and countless other starting scientists by proving that the mammalian embryo had "a huge amount of flexibility," remarks Virginia Papaioannou at Columbia. "Its cells were not determined in what they could do; they responded to their environment."

Mintz discovered that she could combine the cells of as many as ten early mouse embryos into a giant embryo that would revert to a normal-sized embryo when put back into a mouse's uterus. While others would call the animals that resulted from combining two or more embryos together *chimeras*, Mintz prefers the term *mosaic*. The term chimera "reminds me of monsters," she notes.

When she first started gluing mouse embryos together, people thought her experiments would beget exactly that—monsters. But she had no fear of this, she says, because she knew that in the womb, albeit on rare occasions, twin embryos can unite into one embryo, and that, less rarely, one embryo will split into identical twins. And she was right. The chimeric mice she created were completely normal, in spite of their having cells with dissimilar sets of genes. This joining of embryos was the reverse of Hans Driesch's experiment from the 1890s, in which he had shown that an early sea-urchin embryo's separated cells could develop into independent larva. Either way you looked at it, the embryo and its immature cells were extremely malleable. You could add cells to the embryo or subtract cells, and like one of those bottom-heavy plastic dolls you can't bat down, the embryo righted itself to normalcy.

As for why she made chimeric mice in the first place, there was a method to Mintz's apparent madness back then. During graduate school at the University of Iowa, she recounts, she studied "all those classical experiments" whereby Driesch and other experimenters, adding and subtracting cells to and from embryos *ad infinitum*, had used dye and other markers to follow the fate of cells as these embryos grew and developed. No marking method had ever worked very well, "and I thought, oh my gosh, the way to trace what a cell is doing is to exploit its genes as a marker." And that's what a chimeric mouse, being composed of genetically different cells, allowed one to do. To use an inexact example, let's say you add green cells to an embryo that has red cells; by looking retrospectively at patterns established by the green versus red cells as the embryo expands into organized tissues, you could infer the otherwise "invisible developmental history"—Mintz's phrase—of cells differentiating in an embryo. Chimeras also provided a stage for analyzing the earliest cellular beginnings of certain genetic diseases.

The glimpse that her chimeric mice gave her into the fates of cells in the embryo made her tremendously aware, says Mintz, that a "hierarchy of stem cells" was a ruling force in development. A handful of stem cells in the embryo virtually populate and outfit the whole adult. From there on "I realized that the only sensible way to look at development was from the standpoint of stem cells."

When John Gearhart joined Mintz's Philadelphia lab in the fall of '70, Mintz and her coworkers, along with designing chimeras, had begun working with the only embryonic stem cells available from mammals at the time, the embryonal carcinoma cells—or EC cells—that lived in the tumors of Roy Stevens's infamous mouse strain. Relates Gearhart, "There was already a cottage industry of scientists working with EC cells in culture, and it was everyone's goal to see if these cells could be used in chimeras." What seemed invaluable to Mintz would be to exploit the genes in teratoma stem cells as markers as a way of tracking differentiating cells and the tissues they became. By putting teratomas cells from one mouse strain into the early embryo of another mouse strain, you'd have an animal with two sets of genes, and you might follow which cells evolved from which set.

Gearhart acknowledges that once he started working with carcinoma stem cells in Mintz's lab, "I had a little bit of a love affair" with them. "For the very first time ever, we had a pluripotent cell in a dish that could form every other cell type in a dish." It was biology at its basic best. He tried to maintain pure populations of undifferentiated EC cells, yet would admit to defeat. The cells seemed bound and determined to specialize into mature cells; he could never quite hold them back. Still, what an eyeful it was when, in response to some inner drummer, they turned to nerve or muscle or blood or other cells.

In 1972, Gearhart had the chance to meet the man whose research had shoved teratomas and their stem cells into prominence. The occasion was a biology symposium in Venice, and the younger scientist's encounter with Roy Stevens turned out to be even more memorable than it might have been, although not for any scientific revelation.

One evening midway into the conference, a large group of junior and senior scientists had gone out to eat and drink together, ending the night at Harry's Bar, or in that general vicinity. "We got back to the hotel quite late, and a little later there was a knock at my door," Gearhart relates. "Roy was missing. He hadn't come back to the hotel, and they were forming a posse to search for him. He was quite inebriated, and the concern was that he had fallen into a canal." The search party scoured the area—but no Roy. As it happened, Stevens, possibly bored with his cohorts or disenchanted with the conference, on a whim had jumped on a vaporetti, floated up the Grand Canal to the train station, and boarded the Orient Express. "When next heard from, he was in Istanbul," Gearhart recalls.

Stem cells from the tumors of Stevens's mice were so malignant that if they were put into the body cavity of a grown mouse, the mouse would be dead in a matter of weeks. Yet, as Ralph Brinster, Beatrice Mintz, and, in England, Richard Gardner showed in their respective '74-'75 findings, when these tumor stem cells were injected into early embryos, they could enter into normal development and the making of normal tissues. Gearhart had helped grow EC cells for this experiment before leaving Mintz's lab in '75. Indeed, Mintz's was the only team that used genes as markers, which permitted the tracking of cells down specialized pathways.

Here it was two decades later, and Gearhart was trying to corner the same primordial cell in a human fetus that Stevens had shown misbehaved in a mouse fetus and led to those "funny little tumors." Gearhart's reason for pursuing this cell were very different than the reason Thomson and others were pursuing stem cells in IVF embryos. After finishing his postdoctoral training, Gearhart had known this much: He wanted to devote his career to demystifying a mammal's first weeks of development. No other corner of science could possibly be more interesting. Taking a position as a mouse

developmental biologist at the University of Maryland, his knowledge base expanded considerably thanks to the countless evenings he
spent dissecting stillborns with Gladys Wadsworth, a fellow faculty
member and astute anatomist. "She taught me all this—embryo and
fetal anatomy," he recounts with the true ardor of a developmental
biologist. Ten o'clock would arrive, and they might still be leaning
over the open chest of a fetus with Wadsworth in full oration about
the heart and its "great" vessels, or the complicated nerve network
next to the armpit. Recalls Gearhart, "Always the question Gladys
worked with was, What in the very early embryo led to this being
formed this way?"

Gearhart had all the more reason to throw himself into the subject of mammalian development, when, in '79, he was hired by Johns
Hopkins to both teach human embryology and research mammalian
embryogenesis. He pored over moldy embryology tomes and visited
the famous Carnegie collection of pale human embryos suspended
in solution, then housed at the Smithsonian in nearby Washington,
D.C. Meanwhile, "everyone began jumping into transgenics," he recounts, notably transgenic mice. "Transgenic" animals have much in
common with "knockout" animals. Essentially, scientists transfer a
gene from one animal into another. "It became a standard procedure
to try and get at a gene's function this way," says Gearhart.

Gearhart had jumped in, too, hoping to identify genes that drove
the growth of the early embryo. Simultaneously on the side he had a
small project going in which he was analyzing the genes of mice that
had genetic abnormalities similar to those seen in Down's syndrome.
By the early '90s, having spent over a decade on this project and by
then regarded as a first-rate Down's mouse researcher, he was itchy to
extend his investigation beyond mice. "I began thinking, wouldn't it
be nice to study this disorder directly in the human embryo and
heed Alexander Pope's maxim, 'The proper study of mankind is
man.'" That's when he formed the following plan. Down's syndrome
arose in humans who inherit three copies of chromosome 21 instead
of the normal two. He was sure that if he could retrieve stem cells

from human embryos afflicted with the disorder as well as stem cells from normal embryos, then compare how these different versions differentiated, his lab might grasp a tremendous amount about the disease's progression during embryogenesis.

Stem cells in mouse embryos had been isolated; why not those in human embryos? Any other researcher might have faced the stumbling block of obtaining human embryos. But having directed Johns Hopkins's IVF lab for several years during the '80s, Gearhart was pretty sure that some of the couples he had worked with would be more than willing to donate their surplus embryos. When he shared his research plan with David Blake, the medical school's assistant dean for research was afraid that Gearhart might be headed for trouble. A researcher who chose to work with IVF embryos was playing with ethical dynamite. "David's response," recounts Gearhart, "was, 'Is there any other way of getting these cells?' and I said, 'There's Peter Donovan's way—from fetal tissue.'"

Peter Donovan at the National Cancer Institute had recently shown that you could keep primordial germ cells from mouse fetuses growing in culture, as had, in separate work, Brigid Hogan at Vanderbilt University. Both scientists were interested in this cell primarily because it transported an animal's genetic blueprint to the next generation, and not because its counterpart cell in humans might correct disease or assuage aging due to its chromosomes' long telomeres. "Normally these cells last only about a week" in the mouse fetus, notes Michael Shamblott, a developmental biologist in Gearhart's lab. "So the critical milestone" for Donovan and Hogan "was getting them to survive and multiply" in a petri dish, and for a long duration. When these sperm and egg precursor cells did this, they actually converted to a cell that was a phenomenon of culture— an *embryonic germ cell*, or EG cell, as Donovan called it. Unlike the cell it came from, along with dividing endlessly it could give rise to every type cell. What they did to it in a dish actually worked to give it more potential.

Thus, three different kinds of stem cells were now obtainable from mice: EC cells derived from tumors; ES cells freed from embryos; and now EG cells from the forming testis or ovary of fetuses.

Blake's recommendation to Gearhart was that he should obtain stem cells from aborted human fetuses and stay away from the more controversial realm of the embryo. People who saw an IVF embryo as a complete human being would hold a researcher responsible for disassociating its cells and causing its demise. An aborted fetus, on the other hand, had already met its demise within legal limits of the law and apart from any decision made by the researcher. "Researchers are not complicit in the termination decisions," Gearhart today stresses. From a personal standpoint, Gearhart never had had a problem with abortion and fully supported the right of a woman to end a pregnancy. Nevertheless, he would make a thorough sweep of "the inflammatory subject"; he would read countless books on abortion (to this day he starts reading at about three in the morning due to his orphanage-ingrained habit of rising before the birds) and struggle through the mountain of federal, state, and institutional laws and policies that pertain to aborted tissue's research use, not only because he would have to build a strong case for his research to a university review panel, but also for his own peace of mind.

This in-depth review failed to make Gearhart think any differently about abortion. "My shelves are now laden with books and articles, but even after extensive reading and continual discussions, I remain convinced of my position," he maintains. When a fetus is aborted, "a pathologist makes sure that the entire conceptus has been removed from the uterus, and once the pathologist signs off, the fetus is discarded. *Discarded*," he repeats with a shudder. It's the throwing away of fetal tissue that's unethical, he feels, since its cells might translate into lifesaving therapies. He can sound just like Jamie Thomson on the subject of surplus IVF embryos. Thomson told a reporter that using these embryos in research "was not only acceptable, but throwing them out when you could do something useful with them would be unacceptable."

On numerous occasions, Gearhart, his postdoctoral fellow

Michael Shamblott, and John Littlefield, the former chair of pediatrics at Johns Hopkins and one of the fathers of amniocentesis,
huddled together to discuss the related questions of whether to use
stillborn fetuses or fetuses obtained through elective abortions, and
whether to work off private or public funds. "Our primary worry
about using stillborns" and other spontaneously expelled fetuses, says
Shamblott, "was that the tissue might not get to us in fresh enough
condition, and if the tissue wasn't intact, it would be a lot harder to
find the tiny fetal gonad." This minuscule region, wherein lies the
even more minuscule germ cells that they were after, is smaller than a
grain of rice. They therefore decided to use fetuses that were aborted
for therapeutic reasons, which were more likely to reach them in a
fresh state. At the time, this fetus category was eligible for research
funding from the government; however, the Hopkins researchers saw
advantages in opting for the shelter of private funds. Those who
opposed fetal tissue research would not be able to accuse them of
spending so much as a dime of taxpayers' money.

Johns Hopkins's review board gave the go-ahead for the project,
but only after a long and rocky review process that lasted from the
fall of 1993 to the fall of 1996. "It was clear in my mind from the
very beginning that people at Johns Hopkins who had a say in this
really didn't understand the potential of where all this was going,"
voices Gearhart. "The biologists around me got it," but most others
didn't.

A bright spot was that Michael West from Geron materialized in
'96 and clearly did get it. Hearing of Gearhart's bid for the primordial germ cells, the cell that conferred immortality on the human
race, he swooped down on the Johns Hopkins scientist one spring
day as suddenly as he had descended on James Thomson some
months earlier. "He knocked on the door, introduced himself and
his purpose, and took me to lunch," remembers Gearhart. Initially
funded by Johns Hopkins, the stem cell venture soon moved under
Geron's financial wing, although for Geron "the collaboration was
really a hedge," notes West. "I didn't know how easy it would be to

grow [stem] cells from blastocysts," in which case he saw Gearhart's sought-after primordial germ cells as a fallback. One way or the other, whether it came from embryo or fetus, West wanted to have a human immortal cell, complete with its chromosomes' long telomeres.

That's how it ended up that in the fall of 1996, in a warren of subterranean rooms a few buildings away from the medical school's towering Jesus Christ, the statue that greets Christians and non-Christians alike who stream to Johns Hopkins's medical complex from around the world, that Gearhart's lab began working with fetuses supplied by Bayview Medical Center, a Hopkins-affiliated hospital that had an abortion clinic.

Much of the nitty-gritty benchwork was done by Gearhart's postdoc Michael Shamblott. When it came to a hunt for the cells that perpetuate the human race, Shamblott was a perfect accomplice. He had been curious about what living things were made of ever since his uncle, a prominent microbiologist at the University of Minnesota, had performed an experiment for Shamblott when he was nine or ten. "I'm sure that it's why I ended up in science," he relates. It was a typical Minnesota summer's day, as he recalls. The black flies were out in full force, and Gary Litman, his uncle, and Litman's labmates "had rigged up all sorts of creative ways of trapping flies, from squirt guns to fancy swatters." For the boy's enlightenment, Litman put a few of these dead specimens into the chamber of a large lyophilizer—an instrument that sucks all the water out of an object, essentially freeze-drying it—closed the small door, ran the instrument for a few moments, "and in the end pulled out black fly dust," recounts Shamblott. The boy saw that the fly wasn't really much of anything minus its water. "My uncle challenged me and asked, 'What would happen if we added water back to this dust, do you think we'd get a fly again?' I thought and I thought; it was the first time I thought about biology as a question rather than things I had to memorize. We added water back, but only got black paste."

Now, aided by ultra-fine scissors, micro-forceps, and his own steady hand, Shamblott was able to reach into the human fetus and remove its incredibly small genital ridge, the structure that represents the premature testis or ovary. The fetus had to be at just the right stage for primordial germ cells to have migrated into this little area. Any earlier in development, and they wouldn't be there yet; any later, and they might have already begun to differentiate into older, less pluripotent cells. "It was like hitting a moving target," compares Shamblott.

In about a week he and Gearhart had trapped these early manifestations of human egg and sperm cells. There was elation in the lab, but it was short-lived. Placed in culture, the captured cells failed to keep dividing and promoting their continuation outside the fetus in a petri dish clearly would require a better mix of growth factors and feeder cells. For inspiration, they only had to realize that Peter Donovan and Brigid Hogan had gotten the same exact cell from mice to continuously divide in the "outside" world.

It just so happened that at the University of Wisconsin James Thomson was also up against a culture problem. He had known about this obstacle even before receiving approval for his stem cell undertaking in July '95. Both Thomson and Gearhart, by the way, were well aware of what the other was up to and had good reason to be, seeing as how they had so much in common. Both teams were pursuing potent cells that could possibly change medicine so much that the current drug-and-surgery era might seem punishingly barbaric some day. Both were drawing heavily on previous animal experiments to get them there. Both were backed by a futuristic firm whose idea man was either nuts or clever beyond belief. And both were on a collision course with the Pope and about to make religious leaders everywhere run to their scriptures for answers to questions that their forebears had never needed to ask.

The culture problem that threatened to relegate Thomson's embryo research to a back burner had come about because of this scenario: IVF embryos that weren't immediately transferred into a woman were usually frozen right after they were made, on day 1 post-fertilization. If at a later date a woman decided to make use of them for another attempt at pregnancy, they were thawed and grown to day 2 or 3, with transfer into the uterus following. An embryo, however, could not be grown in a dish all the way to day 5 or 6, which in human development was the blastocyst stage, at least not a healthy embryo. And yet that was the stage that Thomson was gunning for, the point at which an embryo's inner cavity forms with its heap of stem cells.

Thomson had gotten Jeffrey Jones, the director of the university's IVF lab, involved, and, according to Jones, they tried, unsuccessfully, any number of ways of improving the culture. "What was so interesting about this project," says Jones, "is that things always came together at the right time." When there seemed no way out of their culture predicament, Jones was at a conference on assisted reproduction in Chicago in May 1996 and nearly fell out of his chair upon hearing an Australian embryologist from Monash University, David Gardner, talk about a culture system he'd invented that kept human embryos thriving and happy straight through to day 5. In the following months, Jones and Gardner discussed the possibility of the Wisconsin team using the Australian's culture approach for its stem cell endeavor, but logistics slowed headway on the matter. Then came another stroke of good fortune. By sheer coincidence, a member of Gardner's lab took a postdoc position at the University of Wisconsin in 1997, and Thomson would contract her to make Gardner's special formula for his project.

What was Gardner's culture secret? He had been trying to better the success rate of implanted IVF embryos—a dismal twenty to twenty-five percent—when he clicked on the obvious. He realized that embryo cultures had been shortsightedly based on the embryo's earliest moments, not later ones. "We learned from mom that she provides different nutrients for different development stages," notes Gardner, now scientific director of the Colorado Center for Reproductive Medicine. So, drawing from vital vitamins and amino acids, he brewed up two media that mimicked the two environments that the embryo sequentially resides in, the fallopian tube and the uterus.

This new approach was so much better than the old one, it would translate into twice as many IVF embryos making it all the way to birth. It also meant the difference between stop and go for Thomson's project. "Before, we could get blastocyst-like embryos, but they had no stem cells in their inner cell mass," describes Jones. "The big change was that the inner cell mass was just huge and beautiful"— and stem cell rich.

As 1997 began, John Gearhart in Baltimore was also beginning to think that his project's culture problem was surmountable. Learning from growth-factor combinations tested by Donovan, Hogan, and others, he and Shamblott were starting to see their embryonic germ cells thrive and last through the weeks, generation after generation.

And then on February 23, out of the village of Roslin, Scotland, swept momentous tidings. For any scientist who might be bearing down on human embryonic cells, the news would have the effect of further muddying waters that had already lost their clarity. Ian Wilmut and his team at the Roslin Institute had done what many scientists imagined was impossible; they had cloned a mammal, specifically a sheep, going about it basically the same way that Briggs and King had cloned their frogs: By putting a cell's nucleus in an egg cell that lulled it back to a totipotent state so that it would start development all over again. Dolly had been cloned from a stored mammary gland cell—thus the researchers' tip-of-the-hat to Dolly Parton—that had been taken from a mature ewe. (Cloning has an odd way of changing relationships. The ewe that gave her genetic information for the making of Dolly can't exactly be considered Dolly's mother, since Dolly wasn't the product of sexual reproduction; instead, the older ewe—who actually was dead when Dolly was cloned from one of her cells—really represents Dolly's identical twin.)

Plant scientists were amused by all the attention given to Dolly, who had been born the previous summer, on July 5. "We said, gee, we've been doing that"—cloning—"in plants for decades, and no one jumped up and down and got excited," recalls Susan Singer. But sheep, unlike plants, were perilously close to humans. Were humans next in line?

The quick succession of cloned mammals after Dolly suggested as much, and the debate over whether or not cloning should serve as a reproductive outlet for the human race raged. Most objectionable to some people was the idea that humans should have this much control over their existence, when we had always been creatures under an immense sky of a higher order. Cloning also raised the questionable, if not frightening, vision of being able to carbon-copy genetically outstanding attributes at will: perfect teeth, perfect IQ, perfect person. The possibility of selecting for genetic qualities suddenly seemed startlingly real. "Dolly really stirred up the whole issue over manipulating an embryo," notes Colin Stewart. The scientific procedure itself was suspect. It took 277 attempts at transferring nuclei from mammary cells into egg cells before the Scottish team had achieved a live birth. The lesson that cloning was neither an easy nor efficient way to create life would continue to be borne out by other scientists.

At the July '97 International Congress of Developmental Biology in Snowbird, Utah, Gearhart announced to attendees of a biology session that he and Shamblott had achieved a first. They'd isolated germ cells from the human fetus and maintained them for several months. The announcement, while causing excitement in the biology world, attracted minimal media coverage. The Johns Hopkins team still had to prove that their cells were pluripotent and further validate their finding by publishing it. By then, the questions and unease precipitated by a sheep's cloning hadn't abated, only intensified, and among the news reports that did cover the Baltimore group's capture of human fetal germ cells, some were symptomatic of the confusion and nervous speculation that had built up post-Dolly. They included "Huxleyish scenarios," to quote *Johns Hopkins Magazine*, that gave the impression that Gearhart was intending to alter genes in his isolated cells, "implant the cells into a woman's womb, and create a genetically engineered person."

The stem cell and cloning details coming at the press and the public weren't easy, and the science was beginning to get all balled up. It was unlikely that either Gearhart's cell from a several-week-old fetus or Thomson's from a six-day-old embryo, if put in the womb, could become a baby. Although both cells have potency, neither cell has a zygote's total potential. But one could appreciate the confusion, especially when so many biologists were working away at modifyng the genes of stem cells, slipping the cells into embryos, and getting changed mice.

Gearhart, that spring, had his lab members read Aldous Huxley's *Brave New World* and thus enter its chilling Center London Hatchery and Conditioning Centre where a "bokanovskified" egg could produce up to ninety-six buds that become ninety-six embryos and then ninety-six identical adults—all identical twins. (The whole process is a bit reminiscent of how hydras bud and produce genetically similar offspring.) The unimaginable was giving way to the imaginable, Gearhart conveyed. Scientists were verging on times when it would be possible to create superior beings akin to the "Alphas" and "Betas" of Huxley's world, and he wanted his postdocs and techs to be good and ready for the hard questions that lay ahead. "If we are successful in genetically manipulating our cells," he says today, "it will mean that now, for the first time, we are instructing our cells to behave as we wish. What enormous powers will we then possess? How will we use them? Who will decide?" The formerly quiet, "recessive" boy, as he remembers himself, was increasingly ready to speak out and stand his ground as a scientist.

Over in Madison, Thomson had to think twice about staying or quitting his course when he first heard about Dolly. "When Dolly happened, I wasn't sure whether I wanted to get caught up in all of this," he recalls. The human embryo was turning into a cause célèbre, and here he'd be dismantling dozens of donated IVF embryos to acquire a cell that so far had no great significance in the eyes of the public. As '97 proceeded, however, the project was spurred along by the chance of growing the embryos to blastocysts due to David Gardner's new-and-improved culture methods. By the year's end the legal language between the American and Australian groups was in order; the project was set to go.

After a long slow drudge, Thomson and Jeffrey Jones finally starting growing their first IVF embryo in Gardner's special media in January 1998. And just as Gardner said would happen, the embryo ripened into a robust blastocyst state. Working in a small room behind the IVF clinic at the University of Wisconsin Hospital, Thomson separated stem cells from this blastocyst—which was the size of a dot, a very small dot—and cultured them as he would have monkey stem cells. From there on in, Thomson, Jones, and those assisting would constantly steal a peek at the cells, hoping and praying they hadn't gotten away from them and differentiated. If the cells weren't split often enough and became too packed together, that could easily happen. At one point, one dish's worth did get by Thomson, their progeny rapidly turning to cardiac muscle cells that beat in unison. "People in the lab were in awe," relates Jones. "It was so symbolic of the therapies these cells stood for."

The weeks ticked by, and the Gearhart group's EG cells and the Thomson group's ES cells kept right on proliferating, generation after generation. *We've crossed the line!* Thomson and his crew realized. They could maintain their cells in culture indefinitely, making for a renewable source of cells that might fulfill the dream of cell-made medicines. Each group wrote up its report, the unavoidable sprint for earliest publication taking place, and in a stunning climax befitting the importance of these comparable coups, the two papers were published within a few days of each other. Thomson and Jones's account of extracting embryonic stem cells from human blastocysts ran in the November 6 issue of *Science*, while Gearhart and Shamblott's report of their EG cells aired in the November 10 issue of the *Proceedings of the National Academy of Sciences*. The Baltimore team had gathered evidence that indeed their fetal cells, like embryo stem cells, could generate cells associated with all three body layers—ectoderm, mesoderm, and endoderm.

Thousands of news groups from around the globe galloped for the story and fastened onto the eye-catching phrase that turned up in both reports: *transplantation therapies.* Cure like with like. Resod grass with grass; renew forest with trees; replace sick cells with well cells. Therapies that relied on stem cells in bone marrow and skin had already been adopted, and without much knowledge of the regenerative cells at the crux of these transplants. Now that researchers knew the value of stem cells, and could make them live and survive and divide outside the body, practitioners of regenerative medicine had their clay, their basic raw material, and could press forward, suggested Gearhart and Thomson, who were not charlatans out to sell snake oil but scientists from eminent establishments speaking through peer-reviewed scientific journals.

But was their hope for the future of medicine too good to be true?

Far away in China, Professor Shunong Li had kept up his progress as well. A colleague of Li's, Bruce Lahn at the University of Chicago, maintains that Li succeeded in isolating stem cells from human embryos and had even kept them alive for five divisions before Thomson's November '98 paper appeared in print. Li could have used some of Roy Stevens's luckiness. According to Lahn, Professor Li lost his cell line when the liquid nitrogen tank it was stored in dried up during the heat of summer. Other seekers of human embryonic stem cells—among them Roger Pedersen at the University of California as well as the Singapore-Australian group of Ariff Bongso, Alan Trounson, Martin Pera, and Ben Reubinoff—were making similar strides, and if Thomson and Gearhart hadn't published when they did, very likely another team or teams would have scored humanity's magical cells a short while later. "The time was right," notes Michael Shamblott.

To no one's astonishment, the twin feats of appropriating stem cells from embryos and from fetuses raised the controversy over embryo research, over abortion, over gene manipulation, and over the question *When Does Life Begin?* to a ferocious new roar. To many Pro-Lifers, Anti-Abortionists, and Born-Again Christians, Thomson's "destruction" of embryos was tantamount to homicide. There were anti-abortionists who considered Gearhart's reliance on aborted fetuses even more egregious since, because of its older age, the fetus was seen to have more intrinsic value and status than an embryo. It didn't necessarily register that Gearhart's research had drawn only on fetuses aborted for therapeutic reasons. Still others "attacked" his work, Gearhart observed in a 1999 letter to Roy Stevens, "on the basis that embryoid bodies"—the little mounds that EG cells piled into if not periodically divided—"were human embryos and that we were making and destroying human beings in culture."

Thomson and Gearhart discussed the possible benefits of human embryonic stem cells with everyone from congressmen to reporters, medical students to drug executives, children to seniors, endlessly. Beyond transplant therapies, these cells might revolutionize other areas, they offered. They could be employed for gauging the effect of drugs on cells and their tissues. To test drugs for heart disease, for instance, researchers couldn't exactly scoop cardiomyocytes from a living heart. Yet conceivably they could transform large batches of pluripotent cells into limitless supplies of myocytes for testing. Additionally, human embryonic stem cells and their behavior in a dish might provide a cinematic view of early development and its explosive process of cell differentiation, right down to the gene level. Genes inside the cells of embryos could be identified, the genes that told a cell, *You're going to be a blood cell,* or *You're going to be a nerve cell.* And grasping normal development was the obvious route to grasping how disease began, and how to thwart it.

As NIH director Harold Varmus summed up to a Senate subcommittee in December 1998 about Thomson's ES cells, "There is almost no realm of medicine that might not be touched by this innovation."

When conveying the promise for medicine, the scientists tried to be realistic. Endless unknowns needed fleshing out. How were stem cells to be expanded into massive amounts? How were they to be uniformly coaxed to differentiate into a cell of one's choice? (Transplanting stem cells themselves was viewed as too dangerous by most scientists, since there was no telling what they might turn into.) Which diseases were the best candidates for transplantation therapy? What was the best way to deliver a batch of mature cells into a patient, and would they integrate with existing cells? How could you be certain there were no stem cells in the batch you were transferring, for fear they would form tumors?

Jamie Thomson, after submitting to intense public scrutiny, would retreat back into his work. Talk, discussion, and explanation, however, would become a way of life for Gearhart. "I decided that I had to become engaged, that I had an obligation to the public to explain the research," he says. His former diffidence gone, after President George W. Bush's August 2001 announcement that federal funds would be available for embryonic stem cell research, but only to researchers working off pre-existing cell lines, Gearhart would be at the front of the charge against this White House decision. It was shocking, he put forth, that the United States, the world's leader in health care, didn't support stem cell research to the hilt. Today, he frequently draws on what Paul Berg, the Nobel-winning biochemist, said to members of Congress in 2003: "Research and demonstrations of clinical efficacy are the only means for validating whether stem cell-mediated therapies will materialize. We are ethically and morally obliged to pursue them for the benefit of those who suffer." Asks Gearhart, "How can you say it any better?"

The ideaman Michael West, whose gave his full support to bringing human embryonic cells to prominence, left Geron early in 1998. He felt held back by the company, and wanted to make better headway toward his goal of reversing aging in cells. Later in '98 he joined Advanced Cell Technology in Worcester, Massachusetts, a company engaged in cloning animals and other futuristic pursuits. As its chairman and CEO, West today is more focused than ever, he says, on research that could reverse aging, and on the embryo's immortal cells that could achieve this breakthrough. He sees therapeutic cloning as the answer. Imagine for a moment that you could start a pre-embryo growing with someone's skin cell, glean its inner stem cells at day 5, and use them to make hemangioblasts—the cells that, lining blood vessels, fall victim to arteriosclerosis, the proverbial "hardening of the arteries."

"Think of it!" exclaims West. "You could re-line that person's arteries. There's a saying in medicine, which is that you are only as old as your blood vessels."

10

Marrow to Brain?

But extraordinary claims require extraordinary evidence.

—Carl Sagan, Astronomer

The sky was robin's-egg blue the spring day in 2001 that I walked down Clarendon Street in Boston's Back Bay neighborhood—past the rusticated stone of Trinity Church, past the sky-high Hancock Tower—arriving in the next block at a brick building that housed the Hard Rock Cafe on its ground floor. "No Drugs or Nuclear Weapons Allowed Inside," the red neon in its window commanded. Upstairs at Viacord, I was to meet with a biochemist who had offered to explain the science behind why a baby's few ounces of cord blood deserve to be preserved and not tossed away like bath water.

Because of stem cells, which were so tiny in size and yet so enormous in their potential, were we on the brink of a truly modern era of medicine? Or were we being led down the garden path by the false presumption that medicines woven from the body's very essence—its cells—might work better against disease than the chemical compounds amassed by the pharmaceutical industry?

One of the very first stem cell scientists I had interviewed was George Daley at the Whitehead Institute in Cambridge. Being an oncologist, Daley had initially approached stem cells from the perspective of cancer. An older view of cancer held that it arose when specialized cells abnormally dedifferentiated back into embryonic or stem cell form, then divided uncontrollably. But opinion was changing, and Daley had aligned himself with the new perception that cancer was a developmental defect. Stem cells themselves might be the problem; if they didn't differentiate when they were supposed to, but kept self-renewing, watch out! It could amount to a malignancy. From there, Daley had gone on to appreciate what these powerful cells might mean for medicine because they *could* differentiate.

Possessing a fiery eloquence, Daley was an inspiring messenger when talking to the public about stem cells. Since stem cells theoretically could give rise to any tissue, give them time and they "could transform the practice of medicine," he had lately told an audience at Boston's Museum of Science. They might lead to therapies for everything from baldness and decayed teeth to the degenerative diseases that had come to represent medicine's "dominant challenge." Cancer, heart disease, brain failure, kidney failure, and other afflictions felt by an aging population "are not easily treated by drugs and their small molecules, *but need cells*," Daley asserted. With an outspokenness admired by his colleagues, he shared his pet peeve with museum listeners, which was the United States government's paltry backing of stem cell research.

A month later, during a press seminar at Whitehead, Daley really got going on the subject. The issue of funding stem cell research had become "a political football," in which the United States would be the loser, he charged. Without the "unfettered" support of the National Institutes of Health, the arrival of valuable stem cell therapies would be held back because hundreds of laboratories couldn't contribute their resources. Already stem cell science in the U.S. was falling into the hands of private companies, which dangerously removed it from public oversight. Furthermore, fired off Daley, other countries, in particular England and Germany, had announced they would support investigations into human embryonic stem cells, putting U.S. scientists "at a significant competitive disadvantage."

Just prior to Daley's Whitehead talk, his son Nicholas had been born, and the newcomer's cord blood had been extracted by his M.D.

dad and subsequently cryopreserved through arrangements made by Viacord. Very possibly, Daley told me a short time later, the stem cell in cord blood possessed nearly the same powers as the bone marrow stem cell that put so many transplant patients on the road to recovery. "You really should go over and talk to the folks at Viacord," he urged.

That spring, the hype over stem cells and their regenerative prowess was at an all-time high. Garrison Keillor even got in the act when a resident of Lake Wobegon began giving shots of stem cell to her tomatoes to make them grow extra large.

Assuredly, a remarkable string of discoveries had just transpired. James Thomson and John Gearhart were still receiving accolades for having collected embryonic stem cells from human embryos and fetuses, respectively, and the press, catching up to the science, was running tantalizing descriptions of what these embryonic cells could be made into: endothelial cells for regenerating damaged blood vessels; osteoblasts for bone repair; islet cells for treating diabetes; kidney cells and liver cells for those respective organs; one type of neuron for Parkinson's disease and another for Huntington's; one type of muscle cell for muscular dystrophy, another for incontinence. At meetings that I attended, researchers traded insights into which growth factors induced ES cells to graduate into a specific mature cell. But clearly this was a difficult art and had a long way to go. The goal of simply "whistling" at ES cells to transform them into a desired cell was still some distance off.

Adult stem cells, the kind that populate the organs of babies and adults, had become a category unto themselves. In fact, just when Thomson's and Gearhart's '98 reports were causing so much commotion, incredible tidings about cells in the adult human brain arrived that fit with the evidence gathered for years from nonvertebrates. Leading up to this news, neural stem cells had been found in rodent brains, confirming Joseph Altman's earlier work. When set in a dish, they morphed into the nervous system's three main cells— neurons, astrocytes, and oligodendrocytes. Experimenters had even shown that when you put neural stem cells back into rodents, they could migrate to a brain site in need of repair. Harvard Medical School's Evan Snyder had injected neural stem cells into the tail vein of newborn "shiver" mice, mice that have tremors because of poor myelin sheathing around the axons of neurons, and the cells journeyed through the blood all the way to the brain, recreated the fatty sheaths, and stopped the shivering of sixty percent of the mice.

As for the news in late '98, a co-effort between scientists at the Salk Institute in California and Sweden's Sahlgrenska University Hospital left sweet evidence that the human hippocampus, the brain's small curled region that figures so prominently in memory and spatial orientation, makes new neurons throughout a person's lifetime. The headlines blazed: The adult human brain regenerated, just like so many other organs. Possibly hundreds of new neurons were added to the human hippocampus every day. Where new neurons lurked, neural stem cells must lurk, and by the new millenium the evidence was irrefutable. The adult human brain contained neural stem cells.

The temptation has existed ever since to correlate the brain's production of new neurons with brain power. Reminiscent of Fernando Nottebohm's canary work, however, scientists have found no easy way to tell if neurogenesis enhances learning and makes a person smarter. The function of new neurons remains undetermined. A study in 2000 revealed that the brains of deceased London taxi drivers had "significantly larger" hippocampi, presumably because of more cells therein, than control subjects who hadn't navigated roads for a living. Can one draw the conclusion that the quantity of neurons in the brain has a bearing on intelligence? Researchers point out that intelligence probably is a lot more complicated.

The new realization that the human brain is capable of regeneration, and isn't the fixed stone of old, has flooded the neuroscience field with optimism. Brain diseases that before seemed untouchable seem approachable now. Cells constantly divide in many regions of the adult human brain; however, according to the current consensus, only in two areas do the daughter cells take the form of neurons: the hippocampus and the walls of the brain's lateral ventricle. These new neurons, and the migratory streams they're part of, might be therapeutically useful. So might the progenitor cells that produce the brain's glial cells. Throughout the brain's white matter, "a vast reservoir of progenitor cells" continuously divides and generates glial cells, notes Steve Goldman at the University of Rochester Medical Center. Exciting Goldman and others is that in a dish these dividing progenitors can be induced to make both glia and neurons. And so a central avenue of research now includes harnessing these in-body progenitors and directing their differentiation so that they might replace dead neurons. White-matter progenitors, for instance, might be a solution for replenishing neurons lost in numerous early-childhood myelin-robbing diseases, "every one worse then the next," Goldman notes.

Researchers in the spring of '01 were beginning to flock to the fledgling field of stem cell biology from every direction. No corner of the sprawling biological and medical sciences was immune to news about these evergreen cells. They had something for everyone, whether hydra specialist, developmental biologist, anatomist, geneticist, biochemist, or pathologist. Sensitive to how easily scientific findings can get blown out of proportion, investigators made gallant attempts to talk about the potential of these flexible cells with level voices. So many scientists were coming to believe in stem cells so fiercely, however, they often unwittingly were a party to the hype.

A few weeks before visiting Viacord, I had attended a seminar at the Dana-Farber Cancer Institute in Boston, where some of the field's top players were lined up to speak, including John Gearhart of EG cell fame and Irving Weissman. Weissman's Stanford lab had persevered in highlighting a stem cell that was the furthest-back ancestor of blood cells ever identified in bone marrow, first in mice in '88 and then in humans a few years later. The meeting's atmosphere in some ways told me more about where stem cell biology was going than did the science discussed that day. Dana-Farber's Jimmy Fund Auditorium was so packed with students and faculty, who lined the walls and even sat on the stage, that scores of others couldn't extend so much as an arm into the room and listened from the corridor.

Once the talks got underway, the audience exuded an intensity that only a crowd of medical mavericks could. On every side of me there was enthrallment, curiosity, and—could it be?—an unspoken spirit of affirmation. When Gearhart forcefully pronounced, "We *can* do this, grow stem cells in certain growth factors and prod them to differentiate," he seemed to be responding to the positive vibes around him, for there seemed to be nothing in that room that day if not faith in the future. I would see this same intense enthusiasm at one stem cell meeting after another; and I would hear it over and over again when speaking with scientists. That day at Dana-Farber, it reminded me of the physician I'd recently heard interviewed on television. Asked for his opinion on stem cells and their treatment value, he had energetically responded, "I have to want to believe in these cells. I took the Hippocratic Oath, didn't I?"

As I climbed the two flights to Viacord, admonishments clung to me nonetheless. Watch out for what you believe, some scientists had warned; hopes for stem cells and their future in medicine could be as overblown as the promise of gene therapy was back in the '70s. About to cross Viacord's threshold into the competitive cord blood industry, I braced for someone in a shiny suit who would wax propagandistic about how the cord blood category of stem cell was free of the controversy that the embryo's stem cell was mired in— even the Pope was behind using it—and about what a waste it was that every year in this country nearly four million umbilical cords got the heave-ho, in spite of their cells' ability to rebuild the blood and immune systems.

But there were no shiny suits to be seen at Viacord. Timothy Moffitt, blond and blue-eyed, wore a wholesome pale yellow button-down cotton shirt and spoke earnestly. To be expected, he recited the advantages of transplanting umbilical cord cells as compared to those in bone marrow. The cord stem cell could treat the very same blood diseases, he said, while at the same time, because it was a more immature cell, it was more compatible with the immune system and posed a lower risk of graft-versus-host disease, the complication that results in a significant number of bone marrow transplant fatalities. Then he patiently responded to my "What abouts?" What about freezing cord blood for decades and suddenly having need of it? Would its stem cells still be viable? Can't say, he admitted, because no one had stored cord blood for longer than twelve years. But there was every reason to think that it could withstand long-term preservation, since the body's blood had been stored for over fifty years. What about the need to multiply cord blood cells into larger quantities for transplanting into adults? Wasn't that important, and wasn't that technology lagging? I'd heard that a child's umbilical cord held only enough stem cells to sufficiently revitalize a child's blood system, but not necessarily an adult's. It was important, he replied, and the technology for expanding the cells was being worked on. I would later learn that the first adults to benefit from this new technology already had received cord blood transplants a year earlier, in 2000.

Although I didn't broach the subject with Moffitt, some physicians believed that families who privately banked a child's cord blood paid too much for too little. Usually there was a one-time service charge (at Viacord, $1,500), along with an annual storage fee in the vicinity of $100—a high price, the critics contended, when there was less than a one-in-one-thousand chance that a stored unit of cord blood would ever be needed. Its stem cells were useful in the face of blood- and possibly bone-related diseases, but that left out many other illnesses associated with other organs, that is, unless scientists were to find a way to dedifferentiate blood stem cells back into embryonic pluripotent cells, the way Frederick Steward had his mature carrot cell, and then get them to redifferentiate into specialized cells for whichever sick tissue. Some scientists would have said that option wasn't even faintly on the horizon. Others would have argued it was.

While companies like Viacord collect and store cord blood for a family's private use, also springing up were public banks that collect a newborn's cord blood free of charge and offer these donated units to anyone in need. A woman from the public blood bank sector told me that processing, testing, and typing one cord of blood can cost as much as $1,400, and that public banks that can't absorb this cost are forced to close. So the private sector's fees seem within reason, especially as a couple can rest assured that a baby's unit of cord blood will be held for that individual or family. Stories circulate of how children have fallen ill, and when their parents have returned to the public bank that took their child's cord blood, they've been told there were no matching units available for their child.

The pertinent question that remains is, how effectively can a baby's banked umbilical cord blood treat an illness that arises in a child or an adult? Data tied to the less tested procedure of transplanting cord blood into adults to revitalize their blood and immune systems look promising. Far more tested and backed by solid data is cord blood's ability to robustly reconstitute the blood-immune systems of children in the event of certain malignant and nonmalignant blood diseases. In some instances, cord blood might prove superior to a bone marrow transplant, since it causes less graft versus host reaction. That said, if the child's sickness arises from an inherited genetic blood aberration, the wisest course of action might be to transplant cord blood from someone unrelated, advises Joanne Kurtzberg, director of the Pediatric Stem-Cell Transplant Program at Duke University. Otherwise, there is the concern that if a child gets his own cord blood cells back, or those of a relative who might have the same genetic defect, the disease could be reestablished, says Kurtzberg. Someday it may be possible to correct genetic abnormalities in the patient's own cells, she adds. For now the main thrust is to infuse healthy cells that can restore lost tissue.

Moffitt, when I saw him in May of '01, said that Viacord had banked in the vicinity of 7,000 cord blood units since its start-up in 1993. The company had been presented with its first case of an allogeneic (one person to another) transplant in June '96; a boy's cord blood had been transplanted into his sister, who had acute lympho-blastic leukemia. Only a month before my visit, in April 2001, Viacord had seen its first autologous transplant. A boy with severe aplastic anemia received back his own unit of cord blood, its cells quickly engrafting and returning him to full health. His particular case having no genetic basis, there was no fear that inserting his cells back into him would recommence the disorder.

As of May 2004, Viacord has banked more than 50,000 cord blood units. A dozen have been utilized – ten for allogeneic trans-plants between siblings and only two for autologous transplants, ac-cording to clinical specialist Kate Falcon. Were Nicholas Daley's blood thawed for use today, there's a far greater likelihood that it would go toward treating a sibling, a parent, or other relative rather than Nicholas himself. Although a healthy sibling's blood may con-tain the genetic aberration responsible for the sick sibling's disease, it also may not, and at the same time prove to be a close genetic match, paving the way for a transplant that is immune-system compatible. There's a twenty-five percent chance that Nicholas's stem cells will be an exact match for a sibling; a fifty-percent chance that it will be a partial match. "This is where the cord blood victory lies," says Fal-con. Because the stem cells from cord blood are more naive than those from bone marrow, there's that much more chance, she says, that they will be an appropriate stem cell source for a sibling without causing an adverse immune reaction. Falcon believes that the cur-rent uses for cord blood are just "scratching the surface," and that in the future the cord blood stem cell will be a fixture in treatments for more and more nongenetic conditions, potentially including every-thing from the rebuilding of spinal cord nerves to the reconstitution of injured cardiac muscle.

After my meeting with Moffitt that day, when I stepped back into the bright sunshine, a sorry sight met my eyes. Seated on a stone wall next to the Hard Rock Cafe was a gaunt, shoeless woman whose lower face was smeared with dirt. Armless, she was smoking a ciga-rette by means of a bare foot. The sight of her sitting there was a coincidental reminder of where stem cells could lead some day. The notion of creating three-dimensional spare parts, whether limbs or organs, may seem more in the realm of science fiction than reality, and yet, as I'd find out, within a short distance scientists at some of Boston's finest research establishments were already using stem and progenitor cells to engineer organs, granted in a rudimentary way. One lab was outfitting dogs with bladders; another was close to creating heart valves for sheep. Other scientists were on the trail of the genes that permit a lobster to regenerate its claw or a fish its fin. Too bad this woman couldn't board a time machine and travel into the distant future, for there's no telling what she'd find—maybe even a new arm.

In the coming weeks and months, stem cells stayed high in the news, due especially to a development that had been heating up since the late '90s. It had to do with adult stem cells, the kind found in an animal's organs. In that they have "committed," as scientists like to say, to a particular tissue, adult stem cells have differentiated a tad compared to stem cells in the embryo, and it was generally assumed that they therefore could only give rise to the specialized cells of their home organ. This was in keeping with the cardinal rule that said that cell differentiation in animals was a one-way street. Cells, like animals themselves, matured, but they couldn't un-mature. They could specialize, but they couldn't un-specialize.

Or could they? Rodent studies and even some human studies were bringing perplexing evidence that adult stem cells perhaps were not restricted in what they could become. Researchers maintained that they had caught sight of *bone marrow* stem cells morphing into brain, muscle, lung, and intestinal cells; *brain* stem cells becoming blood cells; *skin* stem cells turning into muscle and fat cells; *fat* cells turning to bone and muscle. These results flew straight in the face of dogma. "We really had a hard time convincing ourselves of our own data," Angelo Vescovi of the National Neurological Institute in Milan, Italy, told the press when his group presented its finding of neural stem cells turning into blood cells in mice. For this to happen—for mature cells to *transdifferentiate*, or cross lineage boundaries, and become another kind of cell in another tissue—either they had to dedifferentiate and then redifferentiate, or they morphed directly into another cell, which seemed unlikely.

For the cloning of Dolly, the Scottish group had taken a differentiated cell, plunked its nucleus into the right environment—an egg cell—where it had reverted back to immaturity and totipotency, then given rise to a starting embryo that grew into a new sheep. Essentially, then, a mammary gland cell, through the process of transdifferentiation, ended up producing progeny of different lineages of a new individual. Still, most biologists saw this outcome to be the result of a forced circumstance.

Reports by scientists that suggested bone marrow stem cells could transdifferentiate claimed the most attention. And those accounts kept coming and coming. One of the earliest originated in the lab of Éva Mezey, a neuroanatomist with the NIH's National Institute of Neurological Disorders and Stroke (NINDS) in Bethesda. Back in '95, for an investigation into a certain class of brain cell, Mezey had done some bone marrow transplants in mice. Four months after transferring bone marrow from males into young females, she was surprised to notice that some of the females' neurons appeared to be descended from male cells. The telltale evidence was that the nuclei of these neurons contained the Y chromosome, the sex chromosome unique to males. The only conceivable explanation in Mezey's opinion was that, once transplanted, the males' marrow stem cells homed to the females' bone marrow, bore progeny that populated the blood, and then some of those cells traveled to the brain and turned into brain cells. Roughly one to two percent of a female mouse's neurons carried the male chromosome. "That's actually a lot of cells, if you consider how many neurons are in the brain," says Mezey.

Born in Hungary, Mezey had learned from her mentor, renowned anatomist John Szentágothai, that a scientist had to be ready to believe in observations that others didn't have the eyes or courage for. Nevertheless, she was ill-prepared, she says, for the harsh rebukes leveled at her by an extramural board of scientists. "They told me it was irresponsible to come up with unfounded ideas that weren't in the textbooks." Cells didn't jump lineage lanes, and that was that. She sent her report to *Science*, where, after being held for over a year, according to Mezey, it was finally published in December 2000, alongside a similar claim by Helen Blau and coworkers at Stanford. Much to Mezey's relief, Blau had also stumbled on the unusual sighting of bone marrow stem cells switching to nerve cells in mice.

In the wake of these reports, the journal *Cell* published data that were even harder to believe. A collaboration orchestrated by Diane Krause at Yale's school of medicine held out the discovery that one stem cell plucked from an adult mouse's bone marrow had yielded skin, gut, lung, and liver cells. It was the first published demonstration that a single bone marrow stem cell could differentiate so broadly, notes Saul Sharkis at Johns Hopkins, the paper's senior author.

This finding was soon eclipsed by another from the University of Minnesota, Minneapolis. The principal investigator was Catherine Verfaillie, a tall and lanky Belgium-born hematologist who might have become a sports trainer had she not dislocated her knee doing the long jump in school, or a musician had she not broken a couple of fingers also doing school sports, or an engineer had engineering school not called for an entrance exam. And so she had gone to medical school at the Catholic University of Leuven, Belgium, which required no entrance exam. As she conveyed to me in an interview, "Everybody in my family is a perfectionist," and there's no mistaking that she falls into the same basket. She gets to the lab at an ungodly hour, is a disciplined treadmill runner, and admits to being "competitive by nature." During her residency in the mid '80s, her eyes were opened to the great medical value of bone marrow transplants, she recounts, which by then "weren't an outrageous procedure anymore."

Verfaillie's observations arose from some research that she was conducting at the request of Charles Peters, a pediatrician at Minnesota. For some time, Peters had been performing bone marrow transplants in children born with rare genetic disorders, one of which was Hurler syndrome, a skeletal disorder that results from an enzyme deficiency. So severe is Hurler syndrome that unless an afflicted infant receives a bone marrow transplant, their organs swell, they become demented, and they frequently die by age five. A bone marrow transplant can save these children; indeed it is the "only proven effective, long-term treatment" available, says Peters. Yet the young patients are left with terrible bone and joint abnormalities that require endless surgery.

In April of '97, "I went over to meet with Catherine," recollects Peters, "to go over what happens in a successful Hurler's transplant. And basically the question I posed was, were there any ways of enhancing a transplant to help these children so that the vast majority wouldn't have to have significant numbers of orthopedic surgeries?" Because of their surgeries, "to put it bluntly, their quality of life really stinks," says Peters. A journal article had lately reminded scientists of the presence of a second, lesser-known stem cell in bone marrow. This *mesenchymal* stem cell had been identified in the early '90s and was a close cousin to the hematopoietic stem cell. While the latter granddaddy rekindles the blood of bone marrow recipients, the mesenchymal stem cell has the important task of generating the body's bone, cartilage, muscle, and fat cells. (It is named for the embryonic tissue—mesenchyme—from which it originates.) Whenever bone marrow was transplanted, this other stem cell was likely to be present along with its blood-forming cousin.

The idea that Peters and Verfaillie cooked up was that Verfaillie would isolate the bone-making mesenchymal stem cell so that Peters could transplant it directly into his young Hurler patients. "These cells would provide the enzyme missing in these children that breaks down sugars that accumulate in multiple tissues, especially in bone and cartilage, tissues not aided by classical hematopoietic stem cell transplants," explains Verfaillie.

The rest amounts to an accidental discovery. After Verfaillie and her graduate student Morayma Reyes had isolated what Verfaillie believed to be the mesenchymal stem cell, Verfaillie prompted Reyes, who was tending the cells, to omit cow serum from the culture. If and when she and Peters sought FDA approval for their therapeutic method, Verfaillie knew that "the FDA would crack down on using cow serum" in a culture whose cells would be going into children, she notes. So Reyes left out the serum and moved on to plating the cells at a low density—one cell per well, according to Verfaillie. The cells prospered and divided for the next many months, at which point "we had a hunch that our cell was different from your classic mesenchymal stem cell," recalls Verfaillie. For one thing, it grew for much longer. For another, it exhibited a curious idiosyncrasy. One day Verfaillie and Reyes noticed that their cell had differentiated into endothelial cells, which pave the inside of blood vessels. Your "classic" mesenchymal stem cell wasn't supposed to be able to do that.

That was the beginning of a culture odyssey, sending the research into a different orbit. Verfaillie and Reyes made another alteration to the medium their stem cells were floating in, and in their culture dish appeared "strange-looking cells that I vaguely remembered from histology class," recalls Verfaillie. The cells were neurons, oligodendrocytes, and astrocytes, the nervous system's three chief types. "By accident, we had put in all the right ingredients to make neural cells grow." So now they had evidence that their human bone marrow cell—which, lacking some stemness, was likely a progenitor cell—could generate cells indigenous to two layers of the body, mesoderm and ectoderm. At this point, Verfaillie recalls, "on purpose we went after cells from the third layer"—endoderm. And got them! Another culture rendition resulted in liver cells. That their human bone marrow cell could produce such a Noah's Ark of progeny was "absolutely startling"—"a bombshell"—"nothing short of monumental," other scientists stammered to reporters. Verfaillie went back to rodents and showed that if you retrieved the same progenitor cell from their bone marrow and inserted it into the early mouse embryo, its progeny lent to nearly every tissue in the adult mouse. This seemed to strongly validate her cell's multipotent versatility.

Could cells really leap from lineage to lineage? A great many researchers didn't know what to think. Some scientists theorized that, in response to internal emergencies, bone marrow stem cells rode the bloodstream to this or that organ, changing identity when they got there to help repair tissue. This possibility was analyzed and reanalyzed, because if the bloodstream really was a conduit for transdifferentiating cells, scientists might be able to exploit this ferry system as a way of getting more stem cells into sick organs.

Éva Mezey uses multiple sclerosis, a nerve disorder in which the myelin sheath around neural fibers deteriorates, as an example. To try to counterbalance this deficit, the body actually steps up its production of oligodendrocytes, but this attempt can't fix the problem. If transdifferentiation among cells really does occur, Mezey suggests that a treatment option might be to transplant generous amounts of marrow stem cells into the bloodstream that, if luck held, would enter the nervous system, become neural stem cells, and further turn into oligodendrocytes. An even better approach might be to increase the level of growth factors that bolster the body's production of nerve cells. Notes Mezey, "We should let the brain do the job, but help it."

Numerous studies now indicate that when disease strikes, tissues and their cells work even harder to regenerate themselves. Although this repair response is never quite enough to rid a person of the disease, it can restore some function. (Then again, it is possible that the regeneration displayed by so many organs and systems—even the brain—may constantly be keeping illness at bay, and we just aren't aware of it.) A case in point is what happens after a stroke obstructs blood flow to the brain, eliciting tissue damage. In animal models, this insult can trigger the birth of new neurons in the brain's hippocampus as well as around its ventricles. Some of these cells migrate toward the damage and may help in the recovery of neural connections. In humans, this cell surge probably doesn't last longer than a few weeks, yet it may be responsible for the partial recovery that stroke victims frequently experience. Roy Stevens, for instance, has had several strokes since his retirement from the Jackson Laboratory in 1989, and to a certain extent his verbal and motor capabilities have rebounded after each episode. Were stem cells behind this repair, as well they might be, what a just reward for someone who has been so instrumental in bringing stem cells into the public eye.

Conservative groups wasted no time latching onto the idea that stem cells retrieved from adults might be the handmaidens of cell-based therapies. What need was there to destroy IVF embryos, they asked, if cells from bone marrow, fat, skin, or the placenta could be converted to other specialized cells? By the same token, what need was there for human therapeutic cloning, the process of growing an early human embryo to roughly the blastocyst stage so that its stem cells could be harvested and directed toward therapies? No group had accomplished this yet, although someone soon surely would.

Since Dolly, the debate had sizzled on many continents about whether to permit either therapeutic or reproductive human cloning. Therapeutic cloning, which many scientists side with, is strictly a medical strategy. Once born, a person doesn't have embryonic stem cells—only an embryo does—and so this limited cloning procedure can provide someone with their own personal stash of ES cells. One of their cells is cloned, a tiny pre-embryo is advanced to a five-or-six-day-old blastocyst, and stem cells are retrieved that, hypothetically, can be made into specialized cells to treat the person's ailment. A huge plus of "research" cloning, as it's also called, is that the stem cells procured this way are a genetic match to the patient; they wouldn't be rejected by the immune system.

In the case of reproductive cloning, which most scientists vigorously oppose, the blastocyst would be transferred into the womb to produce an infant. This asexual route to human reproduction also had not yet been accomplished. Cloning laws were under review in the United States, where none so far existed at the federal level. In Britain, an added provision to the Human Fertilisation and Embryology Act 1990 made it possible for scientists to therapeutically clone embryos for research, while another ruling had outlawed reproductive cloning.

Starting in '91, members of the Far Right seized upon Catherine Verfaillie's novel bone marrow progenitor cell as the perfect source of cells for transplantation therapies. Verfaillie was less than thrilled. The last thing she wanted was for her research to be hitched onto by people who didn't understand it and were using it to expedite their own agendas; or for adult stem cells to be touted to the exclusion of embryonic stem cells. Even though she had been raised in a devout Catholic family, she was in favor of scientists employing human embryos in research, if their reasons were solid. She had just been made director of the University of Minnesota's new stem cell institute, and she went on record as saying that while she felt that her adult stem cells were capable of extensive plasticity, it was too soon to know how they compared to ES cells, and until both were thoroughly examined, the institute's scientists would "ride the two horses."

The stem cell field meanwhile grew more divided over whether adult stem cells could switch lineages, with the nonbelievers increasing. Out to test transdifferentiation, some groups put marked bone marrow cells into the bloodstream to see if they entered other organs and changed fate. They didn't appear to, at least not to any significant extent. There were sightings, however, of small numbers of circulating marrow cells fusing with native cells in various organs. To many onlookers, cell fusion was more believable than cell transdifferentiation. Egg and sperm cells fused; so did muscle cells. The point was also raised that scientists who had reported transdifferentiation in bench experiments may not have begun with a single cell. That would be the only ironclad way of knowing that the end progeny arose from a single cell.

Catherine Verfaillie today insists that her experiments begin with a single bone marrow cell, which, through proliferation and differentiation, generates an assortment of cell types. Out of the body, her cells, she acknowledges, are probably free of molecular signals that in the body restrict their differentiation. In other words, in a dish her cells might be behaving in ways not possible in vivo. If that's the case, can these cells be therapeutically useful? Verfaillie sees no reason why not. Other investigators aren't convinced that Verfaillie's adult progenitor cell is so outstandingly versatile, although that hasn't stopped Athersys, a Cleveland research house, from attempting to turn this cell into a transplant remedy for genetic diseases.

Verfaillie and Charles Peters still intend to work toward relieving Hurler babies of their skeletal abnormalities. In the meantime, an exciting account coauthored by Duke's Joanne Kurtzberg relates how umbilical cord blood stem cells managed to restore the missing enzyme in the blood and the brain of seventeen of twenty Hurler infants, reversing most of their symptoms. The high success rate had a lot to do with how cord blood from a donor can get along with a recipient's immune system.

The clash of opinions over adult stem cells would ultimately teach me a great deal about cells in general, and stem cells in particular. Just when the reports of cell transdifferentiation were tumbling out fast and furiously, I was spending long hours in the basement of Harvard Medical School's Countway Library, lost in yellowing accounts by early-twentieth-century naturalists about the great lengths they went to in order to try to solve the mystery of how amphibians, crustaceans, worms, and other animals regenerated a missing part, or sometimes even several new selves, as Abraham Trembley had seen hydra do when he cut them into several pieces. (Charles Bonnet agitated over what happens to the soul of a snipped-in-two hydra. "Are there in this Insect . . . as many souls as there are portions of these same Insects which can themselves become perfect Insects?" he had asked.) Early volumes of zoology are crowded with studies on regeneration, the excitement of the chase barely concealed beneath the crust of academic prose. Here's an excerpt from one lengthy report written by Thomas Hunt Morgan that appeared in *The Journal of Experimental Zoölogy* in 1906.

> The experiment was carried out with salamanders collected in the autumn. They were in excellent condition when caught, although not so large as they

soon became when fed on pieces of beef. The individuals were kept for several weeks, without much food, before the experiments began. In some individuals one leg was removed, in others two, in others three, and in still others one, two, or three legs and also the tail, which was cut off near the base. Duplicate sets were prepared, each containing several of these different kinds of individuals. One lot was kept without food and the other fed about every day on small pieces of raw beef. As the new limbs developed they were carefully compared and in some cases measured. It was soon seen that no constant difference could be detected in the two sets, or between different kinds of individuals of the same set, if the regeneration of the new legs is measured by their rate of differentiation. Therefore, food does not seem to be the main factor in the result.

Straining to comprehend how lost appendages were restored, scientists in the early part of the twentieth century were at least one jump ahead of Trembley and his eighteenth-century peers. They knew that the pivotal ingredient in limb replacement were "embryonic cells"—"indifferent" or "undifferentiated" cells. They would detect them in the vicinity of the blastema, the group of cells that form at the border of a missing part. By the 1920s, if not a decade or two before, the regeneration crowd had already reached a surprisingly modern-sounding conclusion about these blastema-related embryonic cells that played into regeneration. Either these immature cells were holdovers from the creature's early development, when it was a mere embryo, or they were the result of mature cells in the adult animal having dedifferentiated back into a pluri- or totipotent state. Either way, a grown creature's new appendage could only arise from embryonic cells; they alone had the potential to produce the array of cell types necessary to reform a lost part.

Interestingly, Catherine Verfaillie has a similar either-or explanation for why she believes her bone marrow cell—found in grown mammals—can versatilely go down multiple lineages. "Either it's a very early stem cell left from development," she observes, "or it's a cell that can dedifferentiate and then be pushed to redifferentiate in multiple directions." What the regeneration literature makes clear is that the stem or progenitor cells in a salamander's stump are mesenchymal in nature—that is, related to the bone-cartilage-muscle lineage—which could be true of Verfaillie's cells. At this juncture, her lab plans to study the genes of its stem cells to see if they bear any resemblance to those involved in a salamander's regrowth ability or that of planarians, microscopic worms.

The evidence nowadays leaves little doubt that the regenerative deftness of hydra, amphibians, worms, and many sea creatures, which Abraham Trembley and his peers were at a loss to explain, is due to mature cells at the injury site backtracking to pluripotency. As such, regeneration incorporates the process of transdifferentiation: Cells of one lineage transdifferentiate into cells of another lineage by first returning to an immature state and then specializing into the diverse cell types needed for forming a new part. Consider what a Max Planck Institute team in Dresden recently witnessed in a salamander's regenerating tail: Spinal cord cells yielded immature cells, and then those pluripotent cells generated muscle and cartilage cells for a new tail. By comparison, hydra are so laden with stem cells that wherever you slice them, as Trembley left evidence of, no dedifferentiation process is necessary; their stem cells are right there, ready to reform whatever section has been lost.

"The ability to regrow missing parts always depends on the availability of a source of pluripotent cells," notes Alejandro Sánchez Alvarado, a biologist at the University of Utah School of Medicine. Which may be why humans are such regenerative laggards, Alvarado and others offer. Our cells and tissues may be so specialized that they don't have the flexibility necessary for regeneration to take place with the same spunk it does in a salamander or zebrafish.

Here the stem cell community was locked in disagreement over whether a *mammal's* cells could cross lineage boundaries, and yet if one stepped back for a broader view, Nature seemed replete with situations in which cells switch fate. It can happen when cells are called upon to grow severed limbs. It can also happen when the eye lens of a salamander is damaged, and pigment cells in the adjacent iris dedifferentiate and redifferentiate to regenerate the lens. It can happen when sponges get pounded to pieces by waves, and mature cells (choancocytes) dedifferentiate, regroup, and differentiate anew to form, by zoologist's Henry Wilson's 1907 account, "perfect sponges." It happens when you take a cutting of a plant, and older tissues regress into embryonic cells that contribute to new root and shoot cells. There are still other ways in which a cell can manifest its flexibility, or plasticity. Put a carcinoma stem cell into an embryo and it can contribute to healthy progeny; or slip a normal embryonic stem cell into a foreign place and it can generate a malignant teratoma.

Dolly the clone is one of the best examples of a cell's inherent flexibility. An udder cell was made to roll back in time, recapture full potential, and launch the development of a new lamb. "Any cell in the body can behave as any other cell. In principle, that's what Dolly is telling us," comments Christopher Potten, a stem cell biologist at EpiStem Ltd., in Manchester, England. *Any cell can become any other cell.* That's because, to use humans as an example, a skin cell on your arm contains the same 30,000 or so genes that are present in a cell in your intestine or a cell in your heart. The difference between cell types amounts to which of a cell's genes are switched on and which are switched off, a difference that is markedly influenced by the cell's surrounding environment. In one environment, a cell's genes will tell it to be a skin cell; in another environment they will tell a cell to be a pancreas cell.

Although it was Dolly who really convinced biologists that the nucleus of a mature mammalian cell had the capacity to dedifferentiate, soon afterward an interesting paper in *Cell* shared its evidence that a mammal's *intact* cell could do the same thing if simply put into a special culture. The Dolly experiment reprogrammed a cell's nucleus by transferring it into an egg environment; here, a cell's nucleus was reprogrammed by changing the cell's entire surroundings. Mark Keating, a cell biologist and cardiologist at Children's Hospital in Boston, and his postdoc Shannon Odelberg took skeletal muscle cells from mice—"the most differentiated cell I could think of," Keating says today—and put them in a solution that inveigled them back into a multipotent state. They then nudged them

to redifferentiate into cells that appeared to be bone, fat, and cartilage cells. "It proved that the term 'terminally differentiated,'" which
refers to mature cells having run their course, "is a misnomer," says
Keating. In both Dolly and this example, genes inside cells were
essentially recalibrated by new environments.

Keating believes that every mature cell in a mammal has this
same flexibility, which could be indispensable for any scientist hoping to direct cells to regenerate diseased tissue. Numerous questions
remain about regeneration, including what starts it, what maintains
it, and what genes account for it. As these molecular enigmas get
solved, "I don't think there's any reason to think that regeneration
can't be enhanced in humans," Keating maintains. "The argument
has shifted from it can't be done, to it can be done."

In his lab above the din of Longwood Avenue, Keating is committed to getting the human heart to regenerate itself. His lab's most
important tool in this respect is the zebrafish, which scientists have
lately realized is as good a regenerator as a salamander. This small
striped fish can renew it fins, spinal cord, and retina, along with a
goodly portion of its heart, Keating and colleagues have shown. Clip
away twenty percent of the heart, and within sixty days it grows back
to its original size and shape, in sharp contrast to the human heart,
which shows hardly any regenerative recovery. The manner in which
the zebrafish heart restores itself resembles the cellular process that
renews an amphibian's appendage: Specialized cells dedifferentiate
and then redifferentiate—in the fish's case, into cardiomyocytes, the
heart's muscle cells. Keating and his labmates are searching for the
genes and proteins that permit this cardiac comeback in the belief
that the human heart has the same capability, although in muted
form. A long-term goal is to inject chemicals that would stimulate
the heart to rebuild lost tissue, says Keating, whose commitment to
the heart ties back to his grandfather's fatal heart attack when Keating
was seven years old.

While disagreements continue among scientists over whether
adult stem cells in mammals have the means to jump tracks and

become a different cell type, the leading vote is that fusion could explain the reports of transdifferentiation, although there are those investigators who maintain that their data show cells crossing lineages—and not fusing. Whatever might be happening in these experiments, the plentiful examples in Nature of cells being reprogrammed and changing direction bring us back to what Matthias Schleiden and Theodor Schwann proposed in the 1830s: A cell appears to have a life of its own.

As little as a cell's flexibility has been explored, scientists are already trying to harness this feature to the best of its advantage for medicine. In 2001-'02, Whitehead Institute colleagues George Daley and Rudolf Jaenisch collaborated on an experiment that illustrated how therapeutic cloning, and its creation of stem cells from five-day blastocysts, might come to the rescue of patients in the future. Their subject was a mouse with a genetic blood disorder. Using a cell from the tip of the mouse's tail, they transferred the nucleus into an egg cell, reprogrammed it, allowed it to develop into a blastocyst, and obtained stem cells. These cells, however, retained the genetic defect, so they corrected the DNA defect, then injected the repaired stem cells into the blood of the original mouse. The cells multiplied and succeeded in ameliorating the mouse's illness. "It was a very exciting outcome," shares Virginia Papaioannou. "There were some problems with the procedure, but it did improve the situation of the mouse. This is what we all say therapeutic cloning can do."

On August 9, 2001, George W. Bush, who had been inaugurated that January, appeared in a televised news conference to announce the White House decision concerning human embryonic stem cells for research. Jordana Lenon, public relations officer at the University of Wisconsin's primate center, recalls that she watched Bush's address in James Thomson's lab surrounded by his postdocs and technicians, although Thomson himself "was reportedly hang gliding," recalls Lenon. Bush informed the public that federal funds would be available to stem cell researchers who utilized any of the sixty-odd human ES cell lines that were already in place worldwide, a number that would shrink considerably due to many of the lines being unavailable. It was a striking declaration insofar as it was coming from a president who was a Born-Again Christian, during times that had an aura of déjà-vu conservatism, in a country whose government had never before sanctioned research on human embryos. Recounts Lenon, "I remember that right after Bush said that, the first thing that anyone in the room said was, 'It's a start.'"

The August decision was neither an executive order nor a law fashioned from judicial or legislative cloth, according to Lana Skirboll, director of NIH's Office of Science Policy. The United States remained strikingly bereft of approved guidelines for research on human embryos that had been created in a dish but never implanted. This absence was so different from the United Kingdom, where, maybe because of Louise Brown, the government had gotten an early jump on regulating IVF embryos for research. In 1984, just six years after the birth of Louise Brown, the Warnock Committee and its chairwoman Dame Mary Warnock had recognized the human embryo's fourteenth day—the onset of the primitive streak—as the very earliest point at which, for research purposes, the embryo began to become a person. This decision set the stage for the UK's 1990 Human Fertilisation and Embryology Act, which allowed IVF embryos to be used for the study of gene abnormalities and diseases, among other indications. An amendment to the act in 2001 had gone further and given researchers license to derive stem cells and cell lines from IVF embryos, and also to create stem cells through therapeutic cloning—with the day 14 cutoff and other regulations strictly enforced.

American stem cell scientists would become increasingly envious of the UK's favorable climate, and yet UK researchers could feel hemmed in by the strict regulations. Any group desiring to use IVF embryos for research "has to jump through hoops," says a member of the Bioscience Unit of the UK's Department for Trade and Industry. "Our policy is very supportive, but very regulative. So it's a typical English stoic."

A little over a month after Bush's announcement, on September 11, 2001, the World Trade Center suffered its mortal blow by terrorists. Making a narrow escape was a fifty-eight-year-old electrician and Parkinson's patient who was working on the thirty-fourth floor of the north tower when it was hit. He had received an experimental transplant of fetal dopamine-producing neurons in January 1999, and a year later had been able to go off the drug Levadopa, according to Curt Freed, the University of Colorado Health Sciences Center neuroscientist who performed the transplant with surgeon Robert Breeze. The man was one of nearly three dozen patients enrolled in an NIH study whose advanced Parkinson's was improved by transplanted fetal tissue, on average by sixty percent, says Freed.

Over the years, the large majority of fetal transplants for Parkinson's have not borne such positive results. Still, the successful exceptions serve as "proof of principle" that dopamine-making neurons, when transplanted in a certain way, can work, agree many scientists. Realistically, fetal cells may never be a good source of tissue, says Freed, because, even if the controversy surrounding fetal tissue didn't exist, there isn't enough fetal tissue for the world's millions of Parkinson's patients. A far better source, believes Freed, might be human embryonic stem cells and their output of an unlimited supply of dopamine neurons.

After running down thirty-three flights of stairs, the Twin Towers survivor walked several miles to the train station. To this day he remains off L-dopa, reports Freed. So far, then, he has twice escaped with his life.

The Art of Cell Replacement

The challenge is to move basic cell biology into real clinical medicine.

—George Daley, Hematologist-Oncologist

Boston's Schepens Eye Research Institute—sandwiched between a nursery school and a methadone clinic—is a Harvard-affiliated research establishment named for the renowned retinal surgeon Charles Schepens. There, Michael Young, a neurobiologist, has carved out a difficult goal for his lab that connects back to the struggles of eighteenth-century naturalist Charles Bonnet.

To briefly revisit that story, Bonnet, his eyesight failing, had attempted to find out if there was any truth in Lazzaro Spallanzani's claim that a salamander could regenerate its eye with the same panache it did its tail. When he tried to remove the eye, Bonnet only succeeded in knocking out its lens and doing other serious optical damage. In a matter of months, however, the salamander's eye would completely heal, new lens and all, while Bonnet's poor vision would steadily worsen.

By now, no one questions Spallanzani's contention. Researchers would just like to know by what means salamanders, newts, frogs, goldfish, and various other lower vertebrates can regrow a lens, iris, or other eye part, so that they might prod the human eye to regenerate in a similar fashion. While this might seem like wishful thinking, evidence already exists that the mammalian eye is responsive to such ploys.

Young and his postdocs have taken on the challenge of regenerating the retina, the thin tissue that lines the back of the eye and converts light to image. Observes Young, research groups around the world have singled out the *re-tin-a*—he pronounces the word with a swift patter of three equally accented syllables—because it is the only major tissue in the global part of the eye for which ophthalmologists lack effective repair methods. By comparison, two other important eye tissues—the lens and cornea—are much more easily treatable. "If the lens is damaged, an artificial lens can replace it," he notes. "And corneal surgeons restore eyesight all the time"—by replacing a defective cornea with a healthy one obtained from a fresh cadaver. Since corneal transplantation's modern beginnings in the '50s, it has become one of the most successful types of transplants available.

But in the case of a diseased or injured retina, few treatment options exist, and consequently the large majority of vision disorders in the Western world are retina related. Among the most common of these are diabetic retinopathy, which affects the eye's blood vessels in diabetics; glaucoma, which affects pressure in the inner eye; retinitis pigmentosa, which affects the retina's epithelial tissue; and macular degeneration, which affects vast numbers of photoreceptors in the retina's center (the macula), causing vision loss in the eye's center.

An affable, low-key fellow in his mid-thirties, Young is a walking—and motorcycle-riding—testament to the extent to which eye afflictions have been successfully dealt with in our modern era. The glasses he wears to correct astigmatism and nearsightedness leave him free for his favorite pastime, a memento of which hangs in his office: a large poster of a bright yellow Ducati motorcycle. If ever there were a time to push further to curb retinal diseases, it is now, stresses Young. A report issued by Prevent Blindness America estimates that as baby boomers age, the number of Americans with impaired vision will double over the next three decades. Macular degeneration alone, which is already the leading cause of blindness in older Americans, is expected to affect one in six individuals over the age of sixty.

Only recently have scientists dared to think that the human

retina might be made to regenerate. It's long been known that retinal cells of certain lower vertebrates are capable of regrowth, but the phenomenon seemed restricted to cold-blooded creatures, whose eyes continue growing in adulthood by adding cells. In contrast, the eyes of warm-blooded mammals grow after birth when cells stretch, not multiply, which made it seem as if there was no cell birth or regeneration happening in the eye, just as there wasn't in the brain. Indeed the retina, which is central nervous system tissue, is an extension of the brain. It was only with Joseph Altman's findings, and then Fernando Nottebohm's, that people began gaining a new perspective on the brains of higher vertebrates, notably bird and mammal. These brains did harbor immature cells that constantly divide and differentiate into specialized cells, and suddenly scientists had a reason to think that the mammalian brain—and the eye—might be candidates for regenerative medicine. Then University of Toronto researchers created a terrific stir when, in 2000, they found stem cells in the eyes of mice, cows, and humans, stem cells that generated retinal cells. The notion came alive that perhaps a mammal's retina did have regenerative power, just like that seen in the eye of Bonnet's salamander, but that some sort of inhibitory process was keeping it under wraps.

At Schepens, Michael Young's game plan is relatively straightforward. After experimenting with and learning from other animals, he hopes eventually to be able to isolate progenitor cells from the human retina, multiply them, and then "transplant them back into the eye and have them become the type of cell we want them to be, which in most cases are photoreceptors cells," says Young.

The human retina contains three main classes of cells—photoreceptor cells, retinal ganglion cells, and inner nuclear layer cells—but of these three, it is mainly the photoreceptors, which consist of rods (for black-and-white and night vision) and cones (for color and daylight vision), that are lost in blinding diseases. (In two notable exceptions—glaucoma and optic neuropathy—the retinal ganglion cells suffer.) Even when photoreceptors are destroyed, for a period of time the rest of the retinal system and its connection to the brain can remain pretty much intact, according to Young, which is all the more reason to think that the strategy of replacing photoreceptors can help people gain back their vision.

Photoreceptors are one of Nature's supreme gifts to animals, and if you have any doubt about this, look at a micrograph of a rod, with its fancy quiver of cilia at one end of an elongated body. Equipped with as many special features as the Mars Rover, a photoreceptor turns light energy entering the eye into chemical energy that gets routed to the brain and is transformed into everything we see. Here's a cell that makes one appreciate the notion that as humans have evolved, our cells may have become too specialized to undergo the changes that a salamander's cells undergo when a salamander needs to replace a part.

When I first met Young in '01, he and his labmates were in the process of extracting progenitor cells from the retinas of healthy baby mice, then injecting these cells into the eyes of mice that had no working rod cells and thus were blind by six months of age. Once transplanted into the space behind the retina, the cells were seen to migrate into the retina photoreceptor region and form what appeared to be working rods. "I think we're pretty confident that they are integrated and have the capacity to function," Young had told me at that point. "But we don't know if they are passing visual information to the mouse." Ascertaining whether a mouse's vision has improved isn't all that easy—the animal can't exactly tell researchers, "Hey, I can see!"—and for this reason, a mouse behaviorist from England was about to visit Young's lab to spend two weeks conducting the necessary tests.

By the time I returned to Schepens nearly two years later, Young and his crew had determined that the mice that had received transplants indeed were seeing a little better, but not because of the implanted cells, exactly. Rather, it was because the new cells that they differentiated into seemed to stop cone cells from degenerating, Young explained. It was a perfect illustration of the maxim, "Science is never what it seems." Young, in the meantime, had formed collaborations with groups in Denmark, Sweden, and California, and was pursuing the same goal as before—retrieving progenitor cells from healthy eyes and transferring them into eyes lacking photoreceptors—this time in pigs. A pig's eye is much closer in size and structure to a human eye than is a mouse eye, and a lot easier to work with, he told me. What's more, "There's an RP pig in North Carolina," Young said with boyish excitement, RP standing for retinitis pigmentosa, a disorder that desecrates rod cells. This pig model was expected to serve as a human-like proving ground for cell therapy.

Throughout the stem cell research community, I was finding, research groups were taking the same course, and still are. In short, they are using animal models that mimic a particular disease as a way of testing the effectiveness of a particular cell therapy approach. Mice with blood deficiencies; rats with spinal cord injury; monkeys with motor atrophy similar to Parkinson's or ALS; zebrafish with heart defects. Learning the biology as they go, these modern-day Bonnets and Trembleys are going for broke. Yet whether cell-based therapies can sufficiently materialize without a more complete understanding of the intricate biology that underlies stem and progenitor cells is one of the questions that researchers are currently asking themselves. Some researchers, fearing that the cart is going before the horse, predict that only when the biology is more fully worked out will effective and safe treatments follow.

During a lecture on stem cells at Schepens in May of 2001, Ruben Adler, an eminent ophthalmologist from Johns Hopkins School of Medicine, said something that stood out both for its refreshing candor and also its reminder that human biology is still at an early hour. Noted Adler, "Everything is so much more complex than one pretends it to be." In his talk—which was titled "Is the potential of stem cells going to be realized?"—he brought home how many things must tie perfectly together if transplanted cells are to integrate with other cells and restore normal function in the central nervous system. Ostensibly, he was talking about cells emptied into the eye. But I would hear other scientists make the same point about cells transplanted into other parts of the central nervous system, whether the substantia nigra, the small midbrain region vandalized by Parkinson's; or the spinal column, the span of, in humans, thirty-three vertebrae that is so vulnerable to injury. Of all the systems that researchers would like to mend with cells, the nervous system is easily the most complicated, and if adding cells can regenerate any part of it, it would be a triumphant endorsement of cell-based medicine in general. In his talk Adler referred to the promising observations centered around using stem cells to restore nerve function, yet warned that the complexity of chemical exchanges that take place between nerve cells and dozens upon dozens of growth factors, which are the proteins that direct and nourish cells, is beyond staggering. "If you're transplanting stem cells into a retina expecting that something within the retina is going to induce their differentiation into the cell types you desire," he told the gathered scientists, "you better pray real hard, because very likely the signaling molecules aren't going to be in the right place at the right time for the stem cells to see."

At a conference on neural stem cells later in 2001, which was held at a Hilton resort next to San Diego's Mission Bay, a neurobiologist I met at lunch one day, Don Gash, had gone further than Adler when he conveyed to me his deep concern about the attempts being made to reverse neurodegenerative diseases through cell replacement measures. Gash, the chairman of anatomy and neurobiology at the University of Kentucky Chandler Medical Center, said that his involvement with early fetal transplants geared to repairing Parkinson's had left him skeptical that a course of action as "simplistic" as putting one type of neuron into a problem area was "going to fix the problem."

Using Parkinson's as an example, Gash cited two major problems. The first had to do with the "monumental" undertaking of trying to reconstruct the complex neural circuitry that Parkinson's pervades. Much of the cellular damage might be concentrated in the substantia nigra, yet neurons in that small midbrain region are part of a loop of processes that include several other parts of the brain. The transplanted neurons "need not only make the right connections, but also receive the right inputs" from neurons in other areas, Gash stressed. I found myself imagining a mechanic trying to fix a car engine by randomly dropping spark plugs under its hood. Those who believe cell transplantation can work to diminish Parkinson's—and that includes many reputable scientists—acknowledge that one of the many details that need clarification is precisely where cells should be introduced in this loop for the greatest, safest effect.

The second problem Gash cited was "control over the neurons you're putting in. Control is a very, very important issue," he warned. If the transplanted cells "do the wrong thing, how are you going to stop them?"

I'd only just been on the phone with a Brigham and Women's neuropathologist, Rebecca Folkerth, who had talked at some length about a disconcerting experience she had had in 1991 that possibly illustrated what Gash was getting at, although no one would ever know for sure. Folkerth, then on the staff of the New England Medical Center in downtown Boston, had been called in one evening to do a brain-only autopsy. The deceased was a middle-aged Parkinson's patient who had traveled to China in 1989, where an American surgeon had implanted fragments of fetal tissues into the putamen region of the man's brain, as well as infused individual cells derived from a six-week embryo into the brain's ventricular system, hoping this surgery would help the patient's severe symptoms. At the time, funding for fetal work of this sort was prohibited in the United States. For many following months, the man's gait, swallowing, speech, and other abilities had notably improved, he and his family had reported. Yet one morning, after telling his wife he felt tired, the man had suddenly died.

That night, when Folkerth dissected the ventricles, she was amazed to find that they were filled with tissue that didn't belong in the brain. "It was bizarre. I could see hair shafts, cartilage, and connective tissue," she recalls. "I was there by myself in the autopsy room, and thought no one would believe me, so I took lots of photographs." The surgeon who had performed the transplant had asked her to check whether the transplanted fetal cells had proliferated. "He really hoped that the surgery had worked to some degree. But I didn't see any nerve cells," she recounts. "My own feeling is that this overgrowth" of nonneural tissue "was compressing his brain stem and that's what caused his death." Where had this extraneous tissue come from? Folkerth imagines that in the course of cutting out and implanting the neural material, the surgeon had unknowingly included progenitor cells from other tissues.

To be sure, this particular case represents a worse-case scenario and emerges from a time, although only a decade ago, when less was understood about brain cell transplantation. It still can serve as a valuable reminder that the very aspect that makes immature cells potentially so useful in medicine—their ability to rapidly multiply and differentiate—can also be their most dangerous.

There isn't a scientist around who would tell you that replacing brain cells is a piece of cake; yet more than a few researchers view the avenue of retinal regeneration as one of the most promising attempts at stem cell therapy currently under investigation. It fits a presumed criterion, which is that stem cells, and the cells derived from them, will be most effective for disorders that are confined to one area of the body and involve one type of cell, and less effective for disorders that are more diffuse and affect many cell types. Juvenile diabetes, the type 1 variety that arises in children, is viewed as a particularly good candidate for cell therapy for that very reason. It involves just one of the pancreas's ten or so cells: the beta cell, which makes insulin. Similarly, Parkinson's attacks dopamine-making neurons. One doesn't hear much discussion about a cell-based therapy for Alzheimer's, on the other hand, largely because the disease creates such a broad swath of destruction in the brain's limbic system. "The problem with Alzheimer's is that whole neural networks consisting of many different types of neurons in various brain regions succumb to the disease," cites Rudolph Tanzi, a Harvard Medical School neuroscientist. "Replacing pockets of cells here and there will not mean that the patient will regain lost connections between nerve cells."

One particularly encouraging achievement for scientists who are working on retinal regeneration is that a stem cell-based therapy for the cornea has already been terrifically successful, and for as long as twenty years. Different from a corneal transplant but often done in conjunction with one, the technique is performed when the surface of the cornea has been lacerated by chemical burns, intense heat, or some other injury. The hero in this case is an adult stem cell that lives in the border, or limbus, of the cornea in the front of the eye, as compared to the stem cell that resides near the retina in the back of the eye. A "limbal stem cell," it's called. Surgeons retrieve limbal stem cells from a patient's good eye, preferably, or from a donor, and then transplant them into the limbus of the damaged eye, where, nearly always engrafting, they replace the dysfunctional cells that had been clouding the cornea's surface and preventing vision.

"The corneal surface almost invariably recovers in ten to fourteen days. The patient can regain total eyesight," notes Kenneth Kenyon, who developed limbal cell transplants in the early '80s during his tenure as the director of the cornea transplant program at the Massachusetts Eye and Ear Infirmary. Kenyon relates that in the '70s one of his mentors at the Eye and Ear, Richard Thoft, had shown that the eye's conjunctiva—the transparent tissue that covers the outer eye—could be transplanted, one eye to another. Scientists are only now realizing that the conjunctiva likely has its own stem cells that permit this grafting procedure. But when Thoft began these transplants, Kenyon points out, "it was before the dawn of stem cells, and no one was thinking stem cells." Thoft's success "planted a seed." With the assistance of Scheffer Tseng, then a research fellow, Kenyon tried his first limbal cell transplant in 1984. It was spectacularly successful, yet initially viewed with suspicion. "Like many a new idea, it took a decade to come into parlance and be accepted."

Patients who had this procedure twenty years ago still show no sign of limbal cell depletion, says Kenyon. "That's the essence of stem cells; they're the gift that keeps on giving."

Another encouraging aspect of research into retinal regeneration is noticeable elsewhere in the stem cell field: For every organ and disease being studied, a range of treatment approaches are under investigation, a situation spawned by scientists' different orientations and strengthened by science's competitive edge. So if one team's approach doesn't pan out, there's the chance that another team's will, or maybe another's. On the Seattle campus of the University of Washington, for example, biologist Thomas Reh is approaching the challenge of replacing cells in the retina in a distinctly different way than Michael Young at Schepens. Instead of transplanting progenitor cells from one animal's eye to another's, Reh would like to goad mature glial cells that are already in the eye, in the vicinity of the retina, to dedifferentiate and then redifferentiate into ganglia cells that would migrate into the retina and replace those killed by glaucoma. This *endogenous* approach of using in-body cells relies heavily on a finding that Reh and teammate Andrew Fischer made a few years back, which is that glial cells in the chick retina have the ability to *de*differentiate and revert back to immature status, then re-mature into a different eye cell. "Since glial cells are spread across the retina, it seems an interesting, sensible strategy—to stimulate glia [in the eye] to undergo this process," justifies Reh.

Compared to the amphibian eye's one-hundred percent ability to regenerate, the chick eye has ten percent regenerative power, and the human eye, zero, says Reh. Why don't birds and mammals regenerate? What's changed in the course of evolution? Those are the questions that presently consume his lab, where the mission is to try to identify factors that will bolster the chick eye's regenerative ability to one-hundred percent. Reh expects it will be many years before researchers are able to tap the human eye's full potential, yet he has implicit faith this will happen some day. Look at the celebrated example of transdifferentiation in the eye of Bonnet's salamander, he exclaims. A salamander's eye lens, iris, and retina can all grow back because adjacent cells, reacting to injury, dedifferentiate and redifferentiate into the appropriate tissues. Most likely, the same molecular components are "still" in place in the human eye, says Reh. What he's indirectly referring to by "still" is that humans were once amphibian, too.

Despite being friends, neither Michael Young nor Tom Reh holds back when touting their respective plans of attack—to transplant cells or to activate in-body cells. These represent the two fundamental approaches throughout the stem cell community. Young at one point said to me, "I hear people say that the job of the endogenous stem cell people is to put transplanters out of business, which makes perfect sense. If you can get the body to fix itself, then you don't have to do a transplant, which is invasive and difficult and fraught with rejection problems. But can it fix itself?" Reh meanwhile thinks that the transplantation approach is equally suspect. "I'm skeptical whether it will work in a highly organized tissue like the retina. I'm skeptical that in transplanting these cells and letting them crawl around, they'll organize so well as to give back vision." He feels, in general, that the history of transplanting cells is "littered with failure." "But hey," he laughs, "I've been surprised every year for the last ten years!"

Whatever research nook one looks into, there is a very real sense that cell-based medicine isn't too good to be true, rather that it's too true from a biological standpoint not to be good medically, and that given time, supportive laws, funding, knowledge, patience, and more time, healing the body with its own cells will be the norm—some day. As researchers make headway in applying stem, progenitor, and even mature cells to diverse disorders, it's expected that society's reliance on chemical-based drugs would diminish, but it's also likely that synthesized chemicals and drugs would work in tandem with cell therapies to alleviate different illnesses.

To date, very few cell-based products exist in the United States. The FDA's Center for Biologics Evaluation and Research [CBER] reports that it has licensed only one cell therapy product—Carticel, which is used to repair cartilage. The Center for Devices and Radiological Health (CDRH) meanwhile has cleared several tissue-engineered products, primarily in the area of skin. (Tissue-engineered products are apt to combine a biological product with a medical device.) FDA approval cannot be taken as a true indicator of the number of cell treatments on the market, because approval is not required for "human cells, tissues, and cellular and tissue-based products that are minimally manipulated" and not combined with a drug or device, according to a CBER spokesman. This noted, the case remains that not many cell products are yet available, and when I made inquiries into why this is, the primary explanation given was that cells haven't been understood well enough to be effectively and safely put back into the body for therapeutic purposes. Also, given that these products contain living cells, the manufacturing and safety challenges are formidable, to say the least.

Despite their optimism, stem cell scientists acknowledge that some of the obstacles facing them are positively gargantuan. Consider a few of the difficulties associated with transplanting cells, a procedure that would first require growing embryonic stem cells, then converting them to a desired cell type, and finally delivering them to the right site in the body. Currently, the reported doubling time of human ES cells is twenty-four to forty-eight hours, depending on the cells; at that rate, say scientists, it could take a very long time to get the therapeutic amounts needed for a procedure. As for coaxing ES cells to become a specific cell type, this remains a difficult art, and so far it's hard to get quantities of almost any desired cell, according to scientists who are trying. Moreover, to what stage of differentiation should cells be grown? Which stage of maturation will be effective for therapy, which will be detrimental? Then there's the question of, when cells are injected, where exactly should they go in an organ? In respect to Parkinson's, for example, some researchers suggest one brain region, while others suggest another. Will enough of the delivered cells live? Will they stay in the right region or will they wander invasively into an area where they don't belong? Finally, if the cells are from a donor, can they be kept from causing a catastrophic immune reaction?

One of the greatest fears is that transplanted cells can create a cancer. Embryonic stem cells are seen as riskier in this respect than older cells, because they are so proliferative. If put in the wrong place, they can lead to teratomas, and perhaps other types of tumors as well, warn scientists. If you're pouring millions of mature cells into the body, can you be certain there's no immature cell among them that will divide uncontrollably and form a malignancy? The endogenous approach—stimulating healthy, in-body cells to replace diseased cells—could be a cause for even greater concern, advises Michael Young. "If you're trying to induce a cell in the body to proliferate that shouldn't be proliferating, the potential to give rise to tumors might be a big problem," he speculates.

A question that is especially hard for researchers to answer at this early stage, yet is so crucial, is, if you go to the trouble of substituting healthy cells for diseased cells, what guarantee is there that the new cells won't suffer the same fate as the old cells? "It probably depends on the context of the disease," says Melissa Carpenter, a stem cell biologist at the Robarts Research Institute in Ontario. "One of the standard answers is that in the case of, say, Parkinson's disease, it might have taken fifty or sixty years for those cells to die." Therefore, the replacement cells might be under no immediate danger.

Scientists are just beginning to understand stem cells; and yet what can come as a surprise is how many therapeutic practices already hinge on the regenerative properties of stem and progenitor cells. Better-known examples include transplants for blood disorders that are reliant on bone marrow, bloodstream, or umbilical cord stem cells; the fusing and healing of bone done by mesenchymal stem cells in bone marrow; and the time-honored skin graft carried out by epidermal stem cells. Lesser known are therapies that utilize blood stem cells to cure immunodeficiency diseases in fetuses; and, as just seen, limbal cell and conjunctival transplants for the eye.

Regenerative medicine is already here, say researchers, due to a very different sector—the pharmaceuticals. Not to be overlooked are the increasing lineup of manufactured drugs that cause stem or progenitor cells in the body to proliferate, resulting in flourishing populations of specialized cells. Examples include erythropoietin, which stimulates blood cell precursors to produce red blood cells and thrombopoietin, which similarly builds up the blood's platelets. Synthesized interleukins and colony-stimulating factors boost the generation of immune system cells. A lab-made parathyroid hormone meanwhile represents the first FDA-approved drug to stimulate bone growth, which it does by causing progenitor cells to make more osteocytes. In respect to the brain, Prozac and other antidepressants reportedly make neurons multiply. Some drugs create the opposite effect and inhibit cell production.

Streaming from the labyrinthian network of biological research in this country and beyond comes an even bigger surprise, which is the impressive number and diversity of cell-dependent treatments that are undergoing rigorous investigation. In the United States, the government's current funding restrictions are said to be curtailing work at the human level and making junior scientists think twice about entering the field. This new science in our midst is such a drawing card to so many researchers, however, it would appear that there's no way of keeping the genie in the bottle. Researchers are finding outlets for studying stem cells at the animal level; or at the human level through private funding; or with federal dollars available for fetal-tissue research or the approved "Presidential" cell lines.

Brief descriptions of some of these research endeavors follow, enough to demonstrate the hungry ingenuity that the notion of treating disease with cells—cells at all stages of maturation—is inspiring.

Be aware that much of this research is in its infancy and must satisfy
many more inquiries into safety and efficacy before gaining acceptance.

MUSCULAR DYSTROPHY

A few years ago, researchers discovered that when bone marrow was transplanted into mice with a skeletal-muscle disorder corresponding to Duchenne muscular dystrophy (DMD), stem cells in the bone marrow traveled into skeletal muscle and actually restored, at least to a small extent, the protein that is deficient in this illness. Observes Emanuela Gussoni at Children's Hospital in Boston, one of the researchers who made this finding, it was an encouraging sign that stem or progenitor cells might be made to reverse the tide of this incurable disease, whose male victims are almost always wheelchair-bound by their mid-teens and often die by their early thirties. For now, researchers face the snag that stem cells in bone marrow don't repair muscle "with high enough efficiency," says Gussoni. Two other cell sources for DMD therapies are under investigation: stem cells derived from muscle, and stem cells derived from the vascular endo thelium of the aorta of embryonic mice. An Italian team recently showed that the latter version, when injected into a leg artery of dystrophic mice, effectively reconstituted skeletal muscles downstream.

AUTOIMMUNE DISORDERS:
LUPUS, MULTIPLE SCLEROSIS

Although different autoimmune disorders affect different parts of the body, their shared feature is that lymphocyte cells in the immune system exert a sustained attack on the body's tissues. In the case of multiple sclerosis, tissue in the central nervous system suffers. In rheumatoid arthritis, connective tissue suffers. In diabetes, pancreatic tissue suffers. In lupus, tissue in the brain, lungs, kidney, and virtually any other organ can suffer.

Peripheral blood and bone marrow transplantation, it's turning out, and the stem cells at their core, just might be the treatment of the future for some autoimmune disorders, if animal and human studies continue to be as encouraging as they have been. For a bone marrow transplant to work, lymphocytes circulating in the blood first are eliminated by chemotherapy and antilymphocyte globulin. Then, when bone marrow or peripheral blood cells are reinfused in a person, it's like "rebooting the computer," likens Ann Traynor, who directs stem cell transplantation for autoimmune diseases at the University of Massachusetts Medical Center in Worcester. Transplanted stem and progenitor cells are counted on to reprogram the immune system, leading to new, naive lymphocytes that "allow tissues to repair themselves." Autologous transplants that accommodate a person's own tissues can be used, thus avoiding the high chance of an immune rejection raised by a donor's tissue. While a person's own stem cells may carry a genetic proclivity for the disease, they are not skewed to enact the disease as are mature lymphocytes, says Traynor.

In a clinical trial currently in progress at Northwestern University in Chicago and at UMass, of the forty lupus patients who have so far received peripheral blood stem cell transplants, seventy-five percent appear in remission from active lupus, some for up to seven years, while twenty-five percent have lapsed back into serious disease, according to Traynor. The NIH, meanwhile, is funding a multicenter clinical trial that will evaluate the effectiveness of bone marrow and peripheral blood transplantation for eighty patients with advanced lupus. If safety and efficacy are confirmed, a larger study will follow.

Transplanting stem cells from bone marrow or the circulating blood may hold promise as well for multiple sclerosis patients, particularly those whose disability is less advanced. "With MS, transplantation appears to be much more effective earlier in the disease than later, after actual nerve loss starts," says Traynor.

Altogether, she believes that better days are coming for autoimmune patients: "Soon we'll see more and more stem cell transplants for lupus, Crohn's disease, and earlier MS. With time, I expect we'll see transplants used for rheumatoid arthritis and diabetes as well. We feel we're approaching a place where people should not die of lupus anymore, should not die of Crohn's disease anymore, should not experience disability that is sustained from these diseases anymore."

TYPE 1 DIABETES (JUVENILE-ONSET DIABETES)

Levels of glucose in our blood are maintained by insulin, and insulin is made by beta cells, one of four types of islet cells in the pancreas. In type 1 diabetes, an autoimmune disorder that appears mostly in childhood, beta cells are largely destroyed, and if the body doesn't get insulin by other means, soaring blood sugar can severely damage the body's organs. This is not a rare disease. Approximately one in every 400 to 500 children and adolescents in the United States has this form of diabetes, estimates the NIH's National Institute of Diabetes and Digestive and Kidney Disease.

Seen as one of the most exciting advances toward liberating patients from insulin injections, a Canadian team at the University of Alberta in Edmonton has devised a procedure—the "Edmonton Protocol"—whereby beta islet cells from fresh cadavers are transplanted, via the portal vein, into the liver. Once engrafted, the cells send insulin to the rest of the body via the bloodstream. Since the first transplant in 1999, the Edmonton group has treated more than sixty patients. "Eighty percent of patients with completed transplants remain insulin-free at one year," according to James Shapiro, the protocol's developer. It's widely observed that the Edmonton Protocol has pretty much proven to the world that if there were enough cadavers, insulin injections might be a thing of the past. Unfortunately, there are only enough cadavers to provide islet cells for a tiny fraction of patients.

Researchers, therefore, are pursuing a better source of beta cells. Since an adult stem cell has not been found in the pancreas, using ES cells to produce beta cells might be the most promising route.

"We know that embryonic stem cells can lead to islets, because if you put an ES cell into a mouse blastocyst, among the cells it contributes to are islets. But we don't know how to do that outside a mouse," says Douglas Melton, a Harvard biologist whose laboratory is working away at this challenge. In 2001, Ronald McKay's NIH group reported that it had actually coaxed a mouse ES cell to differentiate into an insulin-producing cell. A subsequent study, however, has raised some doubt about McKay's findings. So the field goes one step forward, one step back.

Meanwhile, doctors at the University of Cincinnati Pancreatic Disease Center have designed a new cell treatment for pancreatitis that involves removing a patient's diseased pancreas and salvaging as many healthy islets cells as possible; then reinserting the islets into the patient's liver, where, just as in the Edmonton Protocol, they can produce sufficient insulin for the body. Unlike the Edmonton Protocol, patients get back their own islets and therefore aren't beholden to antirejection drugs. The procedure being fairly new, there's no way of yet knowing if the liver-implanted islets will keep working long-term. To date, over forty patients have been treated, all have recovered, and over fifty percent are producing enough of their own insulin. They no longer require injections, according to one of the program's coordinators.

DEAFNESS

Nearly twenty years ago, Douglas Cotanche, a Boston neuroscientist, accidentally discovered that chickens can regenerate hair cells located in their cochleas, the coiled inner-ear tissue so important for hearing. Stem cells in the cochlea appear to account for this ability. According to Cotanche, most hearing loss in humans can be attributed to the loss of sensory hair cells, not to nerve loss, which has spurred his Children's Hospital team to try to induce cochlea stem cells into regenerating sensory hairs in a mammal's ear.

CELLS FOR DRUG TESTING

Convincing embryonic stem cells to differentiate into a specific kind of cell remains a tall order. Yet if the art can be perfected and made reproducible from lab to lab, it could be tremendously useful, and not just for the purpose of transplanting mature cells into ailing organs, but for testing drugs as well. Take the example of a Geron team's having turned human ES cells into what it believes are hepatocytes, the liver's primary specialized cell. If hepatocytes could be mass produced from ES cells, drugmakers, who must show that a new drug has no adverse effects on the liver, would have vast quantities of liver cells. Currently, a chief source of liver cells for drug tests are cadavers.

SALIVARY GLANDS

As patients who receive treatment for cancer of the head and neck know all too well, radiation can cause wicked dry mouth. Explains Simon Tran, a dental clinician at the NIH's National Institute of Dental and Craniofacial Research, "The good news is that radiation can get rid of the cancer, but the bad news is that it can destroy one of two types of cells that make up the salivary glands." To make life a bit more pleasant for patients, Tran would like to replace the cells lost to radiation, even engineer artificial salivary glands from scratch. He therefore has joined up with Éva Mezey at the NIH to see if the bone marrow stem cells that she has proposed can morph into cells of other tissues might serve to replace lost cells in the salivary gland that are responsible for secreting saliva. While there are currently two FDA-approved medications for dry mouth, if a patient lacks saliva-making cells, they aren't of much use, says Tran.

TEETH

Every tooth, it appears, is the appointed outcropping of a few stem cells. Stem cell populations within a tooth give rise to its inner pulp, the surrounding dentin, the outer enamel, and periodontal ligaments that anchor teeth to bone. At the University of Texas Health Science Center at San Antonio, among other places, biologists are studying both the idea of developing teeth directly from human embryonic and adult stem cells, or else triggering an interaction between epithelial and mesenchymal stem cells that might produce a tooth on site. "Our ultimate goal is to be able to regenerate teeth at the site where they are lost," shares Mary MacDougall, a molecular biologist. "I don't think the idea of growing teeth is science fiction; I think it will be science fact." Meanwhile, the dental pulp of baby teeth has been found to be a rich source of stem cells that give rise not only to dentine, but also bone and cartilage.

EGG AND SPERM

Up until recently, it was unimaginable that a mammal's eggs and sperm could be made anywhere but in the body. But in 2003, that quickly changed when a University of Pennsylvania team took mouse embryonic stem cells and differentiated them into eggs in a dish. Not much later, the other shoe fell when investigators elsewhere, including George Daley's Harvard Medical School lab, managed to direct ES cells into sperm cells. Should the derived eggs and sperm check out and prove normal, they could be used to treat infertility in women and men, and refresh the levels of egg and sperm of cancer patients whose own gametes have been clobbered by chemotherapy.

BALDNESS

Stem cells are found in fairly sizable numbers in the bulge of hair follicles in mammals. Their role in this niche is to generate mature cells of the epidermis, sebaceous glands, and follicles. In some cases, researchers are trying to utilize these stem cells for skin replacement; others view them as the answer to hair replacement. Dr. George Cotsarelis's University of Pennsylvania team is pursuing the latter goal and recently took a large step forward. Removing these stem cells from the hair follicle bulge of mice, the Cotsarelis crew transplanted them to furless mice, and in four weeks the so-called nude mice began to grow fur. It's apt to be many years before this same approach can be used to restore human hair. But furrier times may be ahead.

DEPRESSION

Recent studies by Yale researchers suggest that Prozac (fluoxetine) and other antidepressants stimulate the brain's hippocampus to make more new neurons, and that this heightened state of neurogenesis may be why these drugs alleviate depression. Previous studies had shown that brain cell depletion can accompany depression, so it would follow that halting cell loss, as antidepressants appear to do, can reverse it. Some scientists put these findings in a different light, however, arguing that any brain cell changes resulting from antidepressants or other drugs may be harmful to the brain.

EXERCISE AND NEUROGENESIS

In '99, Henriette van Praag at the Salk Institute made a discovery that should have every able-bodied person, young and old, running around the block every day. She showed that when mice were given running wheels and ran to their heart's content, the exercisers, in comparison to mice that had no running wheels and were much more sedentary, made fifty percent more new neurons in the hippocampus, and their ability to learn was significantly greater. In short, exercise may advantageously activate stem cells in the hippocampus. Other studies have demonstrated that exercise influences the brain's metabolic pathways, causing an increase in serotonin and other chemicals. Van Praag notes that to combat depression, "Exercise is probably better than taking antidepressants. If you exercise, you are activating your own system under normal circumstances."

HEART

An oft-repeated phrase in the stem cell community is "the Orlic mouse." It refers to mouse studies conducted by Donald Orlic and colleagues at the NIH and New York Medical College that left dramatic evidence in 2001 that stem cell therapy can fortify damaged heart muscle in mice. The researchers took bone marrow cells that had been partially purified for stem cells, injected these mouse cells into heart tissue, and witnessed a major increase in new myocytes, or muscle cells. The new cells appeared to go to work in the heart, helping the mice to recover and live longer than mice with poor cardiac muscle that hadn't received bone marrow. There was a sixty-eight percent increase in the survival rate, according to Orlic.

Small, intermittent studies with humans leave the impression that stem cells from bone marrow or blood can revitalize human heart tissue as well, although "the qualifier is that these studies must be done in a more rigorous manner," says Orlic. In recent investigations—including two in Germany and one in Hong Kong—upon arriving at the hospital, victims of severe heart attacks received traditional measures of balloon angioplasty and stent implantation or coronary bypass surgery. One week later they also had their own bone marrow withdrawn and injected into the coronary artery responsible for the infarct. Although the benefits varied depending on the patient, on the whole the investigators reported positive results. In some cases, patients showed striking improvement in blood flow and a significant reduction in the zone of damaged tissue. Most significant, all three teams noted that implanting bone marrow stem cells directly into the heart seemed safe.

Stem cells in these situations can't do the impossible; they cannot revive dead heart tissue. Yet it could be that they can generate enough new tissue to give the injured heart a better fighting chance than it would otherwise have. The concept of using stem cells to bolster heart muscle "is brand new in only the last five years," says Orlic. "Before, cardiologists had never been able to form new heart tissue."

INDUCING IMMUNE SYSTEM TOLERANCE IN ORGAN TRANSPLANT PATIENTS

Organ transplantation has come a long way since the first organ— a kidney—was successfully implanted in 1954. Because of Cyclosporine, Prednisone, and other immunosuppressive drugs, as many as ninety percent of transplanted organs survive for at least one year. In dulling the immune system, these drugs can take a toll, however, leaving patients vulnerable to infections, cancers, heart disease, and other conditions. And they often fail to fend off organ rejection in the long run.

Enter a novel procedure that is the talk of the transplant community and one that might mean that organ recipients won't have to rely on these drugs in the future. The procedure's mainstay are immature cells in bone marrow—be they adult stem cells or progenitors. Right before an organ is transplanted, the patient first receives a bone marrow transplant, the marrow donated by the same person who is donating the organ. Immature cells in the donated marrow basically prime the patient's immune system T-cells for the organ that follows, teaching the cells to recognize the donor's cells "as self," says David Sachs, director of the Transplantation Biology Research Center at Massachusetts General Hospital, the first hospital to perform this technique on humans. The first patient to receive the procedure—a kidney recipient—hasn't had to take antirejection drugs since 1998, according to Sachs. If these good results continue, along with freeing patients from the many complications of immunosuppressive drugs, "it could mean longer survival times for organs," predicts Sachs.

BONE

Since there have been bones to break, bones have healed themselves due to the stromal stem cells found within bone marrow. (These stem cells are also called mesenchymal.) These cells produce osteoprogenitor cells, which generate osteoblasts, the mature cells that form bone. Although bone is naturally regenerative, repairing it isn't necessarily simple, as when a fracture doesn't heal, or bone is lost to cancer, infection, a mineral deficiency, a deformity, or an accident. Orthopedic researchers are studying how to control the differentiation of mesenchymal stem cells, so as to use their derivative cells almost as patching ingredients for the skeleton. One approach under investigation is to harvest marrow cells from a large bone with a fair amount of marrow (such as the hip bone), identify the mesenchymal stem cells, expand them in culture, and transplant them beside the fracture or other bone malady. Researchers are also applying stromal stem cells to injured cartilage, muscle, and tendon, which can be harder to treat than bone.

BREAST RECONSTRUCTION

Motivated by the many women who lose a breast to cancer and opt for reconstructive surgery, Hava Avraham, a cell biologist at the Beth Israel Deaconess Medical Center in Boston, is in the preliminary stages of trying to direct stem cells to recreate the mammary fat pad, the tissue that lends structure to the breast. She and two postdocs are looking into whether transplanting either human or mouse bone marrow stem cells or embryonic stem cells into emptied fat pads can lead to tissue replacement. Meanwhile, other groups have reported that there does appear to be a single, self-renewing adult stem cell that gives rise to the human mammary gland's different types of specialized cells.

PARKINSON'S DISEASE

Despite the inherent difficulties of replacing brain cells, advances in growing and managing cells that might be used to treat and alleviate Parkinson's disease invite guarded optimism. For a number of years, Ron McKay has been determining how to bring ES cells through a series of well-defined steps so that they will yield dopamine neurons. "Being able to control transitions—that's what the game is all about," says McKay, who is one of the field's masters at maneuvering stem cells down a specific pathway. In 2000, McKay and his NIH group reported that they had induced mouse ES cells to differentiate into neural precursors, then dopamine neurons. Two years later, they went a step further. After growing dopamine neurons from ES cells, they injected them into the brains of rats with Parkinson's-like symptoms, the rats "clearly benefitting," states McKay.

Apart from the animal work, since the 1980s other researchers have applied experimental treatments directly to patients in the throes of advanced Parkinson's. Neural tissue from aborted human fetuses has been transplanted into disease-affected brain regions of hundreds of patients worldwide, dozens of whom have experienced "up to a fifty percent reduction in their symptoms," according to a report in *Science* in 2000. A few patients have even remained off medication for several years. Researchers see this as an encouraging sign that neural transplants not only can slow Parkinson's but have the potential to reverse its symptoms and sidestep the disease's root cause, which remains unknown. Less encouraging, most Parkinson's patients reportedly have not benefitted long-term from fetal neuronal transplants, and more than a few have developed side effects— notably movement disorders in addition to those they already experienced. While apparently fetal tissue transplantation can sometimes work for Parkinson's, because of its uneven track record, the controversy that surrounds aborted fetuses, and the fact that tissue from as many as six fetuses is needed for every transplant, interest has largely swung to ES cells, and directing them into neurons and transplanting them, as McKay is doing in mice.

Adult stem cells are another possibility. In 2002, Michel Levesque, a neurosurgeon at Cedars-Sinai Medical Center, treated a former fighter jet pilot who has Parkinson's with "remnants of adult neural stem cells" removed from the man's prefrontal cortex, recounts Levesque. The cells were put into the man's left putamen, a disease-affected brain region. The patient has maintained an eighty percent reduction in symptoms on the treated side, according to Levesque, but has developed symptoms on the untreated side. Levesque says he is waiting for the FDA to approve the same surgery for fifteen more patients.

MAKING ORGANS AND OTHER BIO-STRUCTURES

Although using stem, progenitor, or mature cells to fashion organs and other three-dimensional body parts may seem over the top, there's been gradual progress in this area for many years. Tissue engineers follow a basic design. First they create a degradable scaffold—a structure made from a synthetic or natural material meant to support the cells. Then they seed it with cells. While these 3-D creations are largely in the experimental phase, some clinical successes have resulted. One such instance arose in 1998, when a man arrived in the ER unit at the University of Mass Medical School having lost the top of his thumb to a machine. Charles Vacanti, then on the staff at UMass, and his lab crew configured a piece of coral into the same shape as the lost bone and seeded it with the man's own osteopro-genitor cells. They derived these regenerative progenitor cells "from a biopsy of the periosteum (lining) of the man's wrist bone," Vacanti recounts. Once this scaffold was attached to the remaining part of the man's thumb, "the bone cells matured and laid down new bone matrix, filling in the pores of the coral." The man, further reports

Vacanti, "recovered more than ninety percent function in his thumb, as determined by standard hand surgery function tests."

Major organs aren't likely to be produced anytime soon; but there are definite steps in that direction. For example, researchers are employing mesenchymal stem cells from sheep bone marrow to develop heart valves; mesenchymal stem cells from humans to grow bone and cartilage; human vaginal progenitor cells to grow urethras and bladders; ES cells as well as differentiated muscle cells for crafting blood vessels; and liver progenitors to grow liver tissue.

Holding back tissue engineering for many years has been "the inability to expand cells in vitro," according to Anthony Atala, the director of Wake Forest University's new institute for regenerative medicine. It took his lab (previously at Children's Hospital in Boston) as long as five years to learn how to grow specialized urothelial cells, which he went on to use for fashioning entire bladders in dogs. Tissue engineers say that the challenge will be to convert stem cells into large numbers of specialized cells for building specific structures.

SPINAL CORD INJURY

As unthinkable as the notion of repairing the spinal cord once was, it isn't any longer. Researchers believe that in the future the paralysis that presently incapacitates the approximately 2.5 million people living with spinal cord injury worldwide may respond to treatment with regenerative cells. The natural ability of damaged spinal cords of certain adult fish and amphibians to regenerate is taken as an optimistic sign.

Spinal cord regeneration likely will depend on a multifaceted approach. That's because when the spinal cord is severed, "it isn't just that nerves are cut," observes Marie Filbin, a biologist at Hunter College of the City University of New York. Along with neurons, other cells—oligodendrocytes—that myelinate and insulate the axons of nerve cells can die, which also interferes with impulses between the brain and body. Antecedents for spinal cord regeneration possibly exist in humans, and yet degraded myelin will work against these molecular signals. Researchers are therefore headed in many directions. Using rodents, some are directing ES cells in a dish into oligoprogenitors and then injecting those cells into the injury site. Reports cite improved myelination and better locomotion in rats. Other groups are transplanting undifferentiated ES cells directly into rodents to see if they respond to the injury and replace missing tissue. "If you want to repair the [axons] that are cut, instead of getting them to grow a great distance, you might want to transplant in neurons so that they bridge the gap. It seems far-fetched, but it's a possibility," says Filbin.

Filbin's group, in collaboration with Thomas Jessell's at Columbia, is attempting to switch off genes in neural stem cells that inhibit regeneration, hoping that when transplanted, these cells will react to signals to regrow spinal cord tissue. Jessell lately gave transplanters much to cheer about. His lab demonstrated that mouse ES cells could be coaxed to become motor neurons in culture. A Johns Hopkins team then transplanted these cells into the spinal cord, where they survived and even sent out axons toward muscle.

Still other teams are trying to incite cells within the animal to migrate to the injured area of the spine to make new tissue.

AGING

Could aging be closely tied to a slowdown in stem cell production? That's what a small but growing circle of investigators are looking into. A Duke team, for instance, has reported a link between aging in mice and a decline in bone marrow stem cells. With fewer numbers of these regenerators spinning out differentiated cells that repair arteries and keep them healthy and supple, the report related, atherosclerosis developed, a condition in which artery walls lose their elasticity.

Michael Young's lab at the Schepens Eye Research Institute, with its expanding posse of postdocs, has recently moved from a cluster of small, ill-shaped rooms into spacious refurbished quarters down the hall. It's a sign that Young's funding is going well and that the powers that be are warming to the idea of restoring the retina through cell therapy, a notion that just a few years ago was looked upon as overly avant-garde. Throughout the vision field, a contagious interest in cell therapy is inviting a greater variety of studies into how cells might be directed to mend vision woes. Researchers are exploring the use of immature cells from the lacrimal gland to treat dry-eye syndrome; glial cells from the back of the eye to assist in retinal reattachment; immature cells from umbilical cord blood to reconstruct capillaries damaged by retinopathy. One occasionally hears of even more ambitious goals, such as attempts to regenerate a damaged optic nerve or grow an entire cornea, the latter work motivated by a shortage of donor corneas.

What about the ultimate challenge, that of replacing an entire eye? It seems an outrageous expectation. Then again, researchers have identified a patch of cells in the early embryo that becomes the eye, and they can remove these cells from frog embryos and sustain them in culture, whereupon they grow into an eye-like structure. Ask about the future, and it's very much in the eye of the beholder.

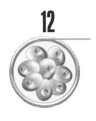

12 After Tomorrow

We are now in the center of biology itself.
—Ronald McKay, Neuroscientist

The scientific field of stem cell biology may not be very old, yet here and there one catches sight of its extreme possibilities. Take, for instance, a study by Japanese scientists that involved planarians, tiny flatworms. Like hydra, planarians are loaded with stem cells, so much so that a little piece of the creature—only 1/279—can regenerate the entire worm, this shown by Thomas Hunt Morgan at the end of the nineteenth century. The recent study had to do with a protein in planarians that normally ensures that only stem cells in the head region differentiate into neural tissue. When the researchers interfered with this protein, stem cells throughout the worm turned into brain tissue. Or, as one planarian scientist observed, "Brain grew all over the worm."

Studies of meristem cells in plants, meanwhile, have biologists envisioning daisies with twice as many petals, lilies with twice as many blooms, and plants that generally might be engineered to perpetually flower. Instead of perennials—maybe diurnals? In the tree department, researchers are endeavoring to discover more about tree meristems and the on-and-off activity of the genes inside them that account for the growth habits of some of the world's most notable specimens. Tipped off by insights into the biology underlying Methuselah, California's 4,770-year-old bristlecone pine, or a redwood's height, or another tree's resistance to disease, growers might produce trees of stunning proportions and hale-and-hardiness. Participants in the Champion Tree Project International already have cloned dozens of outstanding trees by taking cuttings from a niche rich in meristems—the tips of their branches.

For biologists studying animal stem cells, a major limitation continues to be a lack of good markers for identifying different versions of stem cells. Each time the markers improve, new observations result. Earlier this year, a Massachusetts General Hospital team unfurled its surprise finding of stem cells in the ovaries of female mice that had gone unnoticed all this time. Team leader Jonathan Tilly attests that new markers paved the way for this discovery. Should these same adult stem cells turn up in women, the data would topple the old idea that a woman is born with a finite number of egg cells, and present a range of new options, from treating low egg counts to delaying menopause.

As progress into cells and their genes continues, every day the biologist's laboratory becomes that much more of a remarkable place. Every day the ability to isolate, merge, and manipulate Life's little pieces expands. If you went into a Rip-van-Winkle sleep for hundreds of years, what would you find when you awoke? An overly active imagination could have a field day conjuring up a bustling, futuristic street scene, where youthful centenarians (no wrinkles, baldness, arthritis, bad teeth, or age spots) walk their tiger-retrievers and poodle-bears past buildings that are adorned with winter-resistant topiaries (twenty blooms per stem) and manned by plant-animal janitors (four-foot-high dandelion-like creatures), while across the street, in a park filled with lush flora and large tame rabbits (whose fear gene has been knocked out), people fish for giant bass and perch that reproduce so quickly, the pond is never without them.

So much for fantasy. In the here and now, the goals that medical researchers have for stem cells and their derivative cells actually seem within reason, as long as no one expects to find novel cell therapies in the clinic tomorrow. Statements are made that cell-based treatments are oversold, yet what has tended to be "over-hyped," notes biologist Christopher Potten, "is the time scale." The stem cell field as a whole seems aware of this, and more cautious these days about issuing timelines than it did after the uncovering of human embryonic stem cells in 1998. Back then, it wasn't uncommon to hear researchers say that a novel stem cell therapy would be ready in five years. These days, predictions are usually for ten, twenty, or more years, if years are mentioned. At this juncture, stem cell scientists seem genuinely sobered by the formidable chasm of biology to be crossed. Slinging a rope bridge between two Tibetan mountain peaks would be a cinch compared to the herculean task that many researchers have embarked on, which is to identify the "expression" of genes in a cell—which genes are on, which are off—that directs a cell's stage-by-stage differentiation down a specialized path. Possibly as many as ten percent of a cell's genes (in humans, roughly 3,000) undergo changes each time a cell differentiates another stage, Ron McKay reports.

Along with the difficult science, politics and legislation are likely to slow the emergence of stem cell therapies, and already are, in the opinion of many. Worldwide, the debate that has taken center stage is whether to ban both kinds of human cloning—therapeutic (creating blastocysts for culling stem cells) and reproductive (producing a baby)—or just the latter, which, despite persistent rumors, probably has not yet resulted in a live human birth.

Most scientists are stridently opposed to human reproductive cloning. In respect to cloning an animal, close to ninety-nine percent of the attempts purportedly fail, most clones dying soon after implantation. If they reach birth, they can have such terrible abnormalities that, as Rudolf Jaenisch at the Whitehead Institute frequently notes, those that don't survive are "the lucky ones." Jaenisch, who has studied cloning extensively in mice, has found that when a differentiated cell's nucleus is transferred and asked to roll back to a totipotent state, its genes can run amuck, like a clock's inner works when its hands are moved backwards. Scientists, not to mention nonscientists, also oppose reproductive cloning on the grounds that the planet doesn't need such an extreme form of reproduction when there are too many malnourished and neglected children as it is.

Presently, of the several countries that permit culling stem cells from IVF embryos—among them, Singapore, China, Japan, Finland, Greece, Sweden, the United Kingdom, and Korea—a few also permit therapeutic cloning as a way of obtaining stem cells. Other countries, several of which have large Catholic populations and which include Spain, France, Austria, Ireland, Germany, and Italy, oppose either retrieving stem cells from IVF embryos or gaining them through cloning. In the United States, a bill banning both kinds of cloning, therapeutic and reproductive, has passed in the House of Representatives, but similar legislation has been stalemated in the Senate since 2002. The word on the street is that a law that bans reproductive cloning is bound to come into existence, but that similar action against therapeutic cloning would never clear the Senate. As one knowledgeable onlooker told me, "It's unlikely that therapeutic cloning ever will be banned, because too many people in the Senate, including conservatives, have relatives with diseases." At the international level, a group of over sixty science academies has pledged its support for therapeutic cloning. At the national level, several high-profile Republicans, including Arlen Specter, John McCain, Orrin Hatch, and Nancy Reagan, are strongly behind stem cell research, although they don't necessarily agree that embryos should be made for the purpose of gaining stem cells.

Current laws related to stem cell science run helter-skelter in many different directions. In the United States, scientists wanting to investigate embryonic stem cells, and who rely on government funding, have access to a restricted number of human ES cell lines, those made before August 9, 2001. Researchers supported by private funds, meanwhile, can create any number of human ES cell lines from donated surplus IVF embryos, so long as their research is in compliance with state laws. As for researchers who opt to study stem and progenitor cells from either aborted or miscarried fetuses, they face no federal funding restrictions, although they must adhere to certain guidelines that apply to fetal tissue. That's at the federal level. State laws are meanwhile coming into existence that support and promote stem cell research, even to the point of supporting therapeutic cloning (California and New Jersey). Certain other states have laws that prohibit one or both forms of cloning. Therefore, laws in some states pertaining to embryo or fetal tissue research or cloning can run counter to laws in other states, which can run counter to federal laws, which can run counter to legislation in other countries, which could run counter to pronouncements by the United Nations. Currently, the United Nations' 191 members unanimously support a ban on cloning babies but remain at loggerheads over the issue of therapeutic cloning.

An interesting example of a population trying to sort out its feelings toward harvesting stem cells from IVF embryos arose in Singapore in 2001, when a government-appointed bioethics committee began a dialogue with Singapore's lawyers, teachers, engineers, and other professional groups, as well as its numerous religious sects—among them Roman Catholics, Hindus, Bahaists, Taoists, Buddhists, Jews, and Sikhs—asking them for their views. What did each group think of using stem cells in medicine? Were IVF embryos an appropriate source for these cells? Singapore had given its stem cell researchers free rein to utilize IVF embryos for several years, with the closely monitored preconditions that these embryos should not be grown past day 13, day 14 being the start of neural tissue, or used for reproduction. This policy had enabled Ariff Bongso, working with Australian and Israeli researchers, to develop several ES cell lines, putting Singapore at the stem cell forefront. Yet Singapore, home to a sizable Catholic population, had its fair share of citizens who disagreed with the practice of using embryos in research, and before moving ahead with stem cell science any further, the government wanted a better reading of where its citizens stood on the subject.

The letters written to Singapore's bioethics committee, which can be read on the Internet, render a moving portrayal of religious leaders searching through their ancient books, and their souls, for guidance in making the right decision. Several responded that no-where does it say what to do in this case. It's a painful depiction of religion trying to keep pace with science, which, very plainly, it was never designed to do. With the exception of a few groups, including the Catholic Church and the Sikhs, the majority of respondents said that they accepted the practice of harvesting stem cells from em-bryos, agreeing that day 14 seemed an appropriate cutoff point.

A question that gets asked a lot is, since a fair number of people object to using embryos for stem cell research on strong moral grounds, why don't scientists make do with the adult stem cells that sparsely populate organs? An inescapable truth is that the adult ver-sion are the only human stem cells so far employed for therapies in humans. Stem cells from embryos, or mature cells grown in culture from them, have yet to successfully treat a human disorder. When pressed on this, investigators are apt to respond that human embry-onic stem cells have only just been isolated. "Give us time to figure them out!" They also point out that twenty-five years elapsed from when embryonic stem cells were isolated from mice to when mean-ingful experiments began to occur with those cells. The mouse suc-cesses, while limited just a short time ago, have come on strong of late. There are early indications in mice—which should be seen as a long shot from humans—that mouse ES cells or their mature de-scendants can improve, to varying degrees, heart and stroke condi-tions, neurologic disorders including spinal cord injury, a range of blood and immunodeficiency disorders, bone irregularities, diabe-tes, and still other abnormalities in mice.

The ability of embryonic stem cells to give rise to a greater circle of specialized cells than the adult version is seen as their number-one drawing card. As younger cells, their immortal, self-renewing ways also avail them and their progeny of being much more long-lasting, whether in a dish or in the body, than adult stem cells, whose ability to divide and differentiate can start to diminish. On the other hand, the proliferativeness of embryonic stem cells might make them more prone to causing a malignancy than the adult kind.

The hunt meanwhile continues for a useful source of human stem cells that everyone might be behind. Over the last couple of years, several different adult stem cells have been pushed into the news by the suggestion that here's a stem cell—whether from bone marrow, umbilical cord blood, amniotic fluid, body fat, skin, baby teeth, cadaver, or other source—that won't cause any moral and ethical upset, and perhaps could be as effective and ambidextrous a cell as the stem cell from embryos. However, especially since adult stem cells have taken a step toward differentiating, many scientists are reluctant to put faith in the notion, before hard evidence arrives, that adult stem cells have the versatility and longevity that an embryonic stem cell can offer.

As for embryonic stem cells, there are more sources available than immediately meet the eye. They can be obtained from IVF embryos. They can also be arrived at through therapeutic cloning, whereby a donated nucleus is inserted into an egg cell, grown to blastocyst stage, and the blastocyst is harvested. In February '04, a South Korean team reported that they had achieved the inevitable and procured a continuously dividing line of human embryonic stem cells this way.

Embryonic stem cells can also be derived from a human parthenote, an unfertilized egg cell that is stimulated, by chemical or electrical impulse, to start growing like an embryo. It, too, can be grown to roughly blastocyst stage, its stem cells then harvested. Certain species of insects, fish, and lizards reproduce parthenogenetically, with an egg launching into development without a sperm. But it's only lately that scientists have gotten mammalian eggs to take this track. A parthenote allegedly cannot advance to fetal stage, since it lacks paternal DNA that promotes the growth of the umbilical cord. Some scientists reason that because a parthenote is not a product of egg and sperm, it can't be considered an embryo, making it a noncontroversial source of stem cells. Whether a parthenote's stem cells function normally still needs validation.

From China comes news of yet another way to derive embryonic cells. A former NIH researcher reports taking human skin cells, transferring their nuclei into rabbit eggs, and, through this inventive instance of cloning, growing human embryos to five-day blastocyst stage to retrieve their stem cells. A major advantage of this approach is that it would avoid using human eggs.

All told, any method of obtaining human embryonic stem cells can be looked upon as morally objectionable, since it involves growing up human life—whether an IVF embryo, an embryo produced by cloning a cell, or even an egg cell—to roughly day 4 or 5 for its inner bounty of stem cells. A handful of groups have recently claimed, however, that their experiments have bypassed the embryo and its stem cells altogether. They maintain that they have developed culture techniques for turning one type of mature cell into another type. This represents transdifferentiation, the process so often disputed. The British company TriStem in London, for example, maintains that it can extract a person's white blood cells, dedifferentiate them to a stem cell-like state, and then direct them into other specialized cells. Timothy McCaffrey, a biochemist at George Washington University in Washington, D.C., tested the method in his lab and calls it "striking" and "an ethicist's dream." "You can take blood from a person in the morning, and by the afternoon those cells have reverted to a primitive state. Then you manipulate them into cardiomyocytes or neurons," says McCaffrey. If the method bears out in human trials, its greatest benefit, he believes, would be to allow patients to receive back their own cells as treatment for a disease instead of someone else's, which can set off the immune system.

More than a few scientists voice skepticism over a technique that so easily changes one cell to another. "If cells can run around and become any other cell, why aren't I a pile of goo?" observes one stem cell scientist. Other researchers mention that even if a mature cell could be demoted to immature and made pluripotent, it would contain mutations that typically accumulate in the DNA of older cells, making it and its progeny less than ideal for cell therapy.

Still, cloning and its ability to revert the nucleus of a cell to a pluripotent state make even veteran biologists believe that, in theory at least, it should be possible to take any cell, sit it in the right culture, and prompt its genes to revert back to a zygotic state capable of beginning an embryo, leading to a person. Then the question comes down to, if every cell has the ability to become a human being, does that make every cell sacred? Several years ago, Harold Varmus, who currently presides over the Memorial Sloan-Kettering Cancer Center, was quoted in *The New Republic* as saying, "If we say any cell has the potential to be a human being, then every time you cut your finger, do you have to wear black?" Most stem cell scientists as well are pretty clear about where things start crossing over the line into absurdity. "The Dolly experiment says that any cell in your body is totipotent, but that doesn't make every cell in your body equivalent to a baby," maintains James Thomson.

Investigators estimate that for decent progress to be made in developing cell therapies, hundreds, if not thousands, of human ES cell lines are needed worldwide. In the United States, of the seventy-eight "presidential lines" that qualify for federal grants, as of May '04 less then twenty are usable. They include three batches resulting from James Thomson's work at the University of Wisconsin and five from Ariff Bongo and colleagues in Singapore. Even if all seventy-eight were up for grabs, none could go toward patient treatment, since they were made with mouse feeder cells, which can carry viruses. In 2002, Doug Melton, then chairman of Harvard's Department of Molecular and Cellular Biology, told me that "it would be surprising if the first lines were any good" in the first place. "If you look at the mouse embryonic stem cell work, it took years before robust cell lines were established," many of the early ones having been "duds." Melton estimated that at the time at least ninety percent of stem cell investigators were doing basic research with animal stem cells while they waited for the political scene surrounding human ES cells "to sort itself out."

Melton didn't want to wait, not after having learned first that his son had juvenile diabetes, then that his daughter had the same disorder. Teaming up with Doug Powers at Boston IVF and Andrew McMahon, another Harvard biologist, he put into motion the gutsy plan of creating human ES cell lines using frozen IVF embryos contributed by couples. Since he had generous private funds obtained through the Howard Hughes Foundation and the Juvenile Diabetes Research Foundation, he was not beholden to federal restrictions and could proceed. As for state restrictions, Massachusetts had passed a law in 1974 that banned research on live human fetuses, which, by extension, could be construed to include IVF embryos. Yet a researcher could be exempted from this ruling by getting an institutional review board's approval for a research undertaking, which Melton got from Harvard. And so, by intelligently dodging bullets, Melton was in the clear.

Boston IVF began thawing surplus embryos for the project in January '02. They were transported to Melton's lab in Cambridge, where Melton and a team of assistants grew them to blastocyst stage, removed the blastocysts' outer cells, set loose their inner pearls, and cultured them. When I visited later in '02, Melton said he had at least two, maybe four, "non-presidential lines" already flourishing. As soon as he made a few more and verified their pluripotency, he planned to start giving them away to research groups, free for the asking, in comparison to the cell lines that were eligible for federal grants, which cost up to $5,000 per line. "I can't wait to give them out!" he exclaimed. He saw each cell batch as an experimental treasure for medical researchers. The unfortunate hitch was that, in the U.S., only researchers with private funding would be able to profit scientifically from these new lines. Researchers outside the country might benefit to a greater extent.

Now, in the spring of 2004, Melton has as many as seventeen human ES cell lines in the fridge and is just beginning to send them off to other benches. Meanwhile, at Boston IVF Doug Powers reports that as many as thirty to forty percent of couples who have finished their infertility treatment are choosing to donate their extra embryos to the stem cell project.

From across the country, and from other continents as well, including the UK, Singapore, Australia, and Europe, comes news that a growing number of laboratories are bringing fresh new ES cell lines into existence. The freshness factor might be an important criterion for their use, for it's been discovered that ES cells that grow in culture for long periods can acquire mutations in their chromosomes, which could possibly mar their clinical value. "There may be an ongoing need to generate new ES cells," says Andy McMahon. "But we won't know the answer until we work on more lines." Melton's stem cell lines, like so many others, have been grown in the company of mouse cells. McMahon doesn't see that as a hindrance at this early stage, as there's much more to learn about ES cells before they take the form of new medicines.

And so, ingenuity by ingenuity, the stem cell field presses onward—not quickly, but with an eagerness that neither difficult science nor unsettled legislation can necessarily squelch. As young as the stem cell field is, already there are signs of certain obstacles giving way. For example, Ariff Bongso's Singapore lab and other groups are developing ways of culturing ES cells without animal feeder cells, which otherwise would keep ES cells from clinical use. Also, the technology of rendering pure batches of stem cells continues to improve. Stanford's Irv Weissman believes that if you can produce a completely pure population of hematopoietic stem cells, one that excludes the mature lymphocyte cells that trigger a person's immune system, it would be possible to transfer bone marrow stem cells from one person to another without having to worry that the transplant would cause an immune system rejection. "Because there would be no T-cells there to trigger one!" exclaims Weissman. This sort of creative problem solving is everywhere to be seen in the field.

An area that is turning into a research hot spot is the study of stem cells, not as a treatment for disease, but as the origin of disease. Since stem cells are biological building blocks, presumably if they falter, they carry the recipe for disaster. NIH bone specialist Pam Robey refers to her lab's cells, which are bone marrow stromal (or mesenchymal) stem cells that give rise to bone, cartilage, and fat, as "the cause and the eventual cure" of the rare genetic skeletal disorders that she and her labmates investigate. "Any mutation, any environmental factor that causes a change in stem cell metabolism can lead to one of these diseases," she says. "The cure, meanwhile, is skeletal regeneration, and figuring out how to incorporate our stem cells into the skeleton."

In the area of cancer, to what extent different cancers are the result of stem cells that go awry, no one can yet say. However, it's easy enough to see how stem cells can launch a rapidly growing malignancy. Notes Christopher Potten, "If stem cells are normally dividing in a 50-50 way," where fifty percent are self-renewing and fifty percent are differentiating, "to get cancer, there would only need to be a change to 49-51," or a slight shift to too many self-renewing cells. Whether cancer really can be called a stem cell disease, which implies that the stem cell is wholly responsible, needs further examination. This said, various researchers have recently presented evidence that tumor-inducing stem cells may play a role in breast cancer, gliomas (the most common kind of brain tumor in adults), and a variety of leukemias.

In Harvard's Countway library hangs Robert Hinckley's oil painting, *First Operation Under Ether*, featuring doctors in somber dress and a patient out cold with his head thrown back. The occasion, which took place under Massachusetts General Hospital's Ether Dome in 1846, was the first public demonstration of ether's effects. It's a large, heroic painting, capturing as it does the challenge of curing people in a humane way. To this day anesthesia remains a godsend. Hinckley's portrait, though, seems set back in the Dark Ages. Not even ten years had passed since cells had been recognized as the little worlds that Life is made of. One hundred and fifty years later, are we finally at a turning point of knowing enough about cells to begin to make them our medicines? Is it just an illusion that medicine can be better, and at the same time kinder, than what patients in the past have experienced? The new presence of stem cells in our midst opens a new chapter and makes one a believer.

Some biologists believe that the greatest legacy of stem cells will lie in their therapeutic value. Others say their best gift will be what they can tell us about the miracle of Life, and how a tiny cell has the wherewithal to grow into a thriving form, of whatever nature. Other biologists instead look at a stem cell as the immortal cell that will extend human life indefinitely and reveal aging to be a mere aberration. That may sound far-fetched, but it isn't altogether. One of Abraham Trembley's successors, Howard Lenhoff, tells of how, when he was just starting his career, he fed the same hydra for three years, until he got "bored with feeding it" and stopped. Had he kept up the routine, the animal would still be living, he maintains. "A hydra's cells have a life span of about six weeks; they constantly die and are sloughed off." And yet at the same time, "they are constantly being renewed."

"You see," he says, "a hydra is essentially an immortal animal. It doesn't die."

Notes

INTRODUCTION

Source material includes interviews by the author with George Daley: May 29, 2001; June 21, 2001; August 16, 2002; October 3, 2003; January 19, 2004. Kenneth Kenyon: November 1, 2003; January 16, 2004. William Lensch: August 16, 2002; October 8, 2002; August 21, 2003. Susan Singer: April, 10, 2002; May 16, 2002. Richard Vulliet: March 22, 2004.

    p. 3 **"The essence of a stem cell":** Kenneth Kenyon, interview by the author, November 1, 2003.

    pp. 3-4 **But each of these cells is much tinier:** One plant seed, in fact, can contain hundreds of thousands of stem cells.

    p. 4 **that a sperm from his "aged, decrepit body":** George Daley, "Stem Cells: Medicine and Myth," "New Horizons in the Post-Genome World," press conference, Whitehead Institute, May 4, 2001.

    p. 6 **to restore a full complement of red blood cells:** One such study: M. Osawa . . . H. Nakauchi, "Long-term lymphohematopoietic reconstitution by a single CD34-low/negative hematopoietic stem cell," *Science*, Vol. 273 (1996), pp. 242-45.

    p. 7 **by a bipartisan group of U.S. senators:** Sam Brownback (R-KS), Orrin Hatch (R-UT), and Chris Dodd (D-CT) announced the proposal of the Cord Blood Stem Cell Act of 2003 in the fall of '03. (The act contains two bills—S1717 and HR 2852—the latter bill previously proposed by Christopher Smith [R-NJ]). Congress has not yet reached a decision. For an update, or to read both bills, go to www.thomas.loc.gov.

    p. 9 **"Even the dumbest stem cell is smarter":** Evan Snyder, interviewed by Carl T. Hall, "Harvard researcher mines stem cell riches in West," *The San Francisco Chronicle*, January 6, 2003, p. A6.

p. 9 **Similar to the animal kind, meristem cells:** The numerous biological elements that plants and animals have in common, which include stem cells, lead many scientists to believe that plants and animals are descended from a common ancestor. Embryologist Scott Gilbert comments in a footnote in *Developmental Biology* (6th edition, p. 635): "The similarities between plant meristem cells and animal stem cells may extend to the molecular level, indicating that stem cells existed before plants and animals pursued separate phylogenetic pathways."

p. 9 **"If you go into a field":** Susan Singer, interview by the author, April 10, 2002.

p. 10 **"it brings to mind amphibians":** Biologist from Novartis, conversation with the author, June 9, 2003. Several other scientists made similar comments to the author, among them Mark Keating, who studies amphibian regeneration: "I think for a long time regeneration has been in the badland of science." Keating, interview by the author, May 14, 2003.

p. 12 **Stem cells injected into the coronary artery of dogs:** P.R. Vulliet . . . M.D. Kittleson, "Intra-coronary arterial injection of mesenchymal stromal cells and microinfarction in dogs," *Lancet*, Vol. 363 (2004), pp. 783-84.

1
PLANT OR ANIMAL?

Source material includes interviews by the author with Hans Bode: March 11, 2002; March 12, 2002. Richard Campbell: October 2002. Charles Dinsmore: March 15, 2002. Howard Lenhoff: February 9, 2002; May 5, 2002; June 10, 2002.

p. 13 **"If there were no regeneration":** Richard J. Goss, *Principles of Regeneration* (New York: Academic Press, 1969), p. 1.

p. 13 **Many Junes ago:** Howard and Sylvia Lenhoff and their wonderfully written and elegantly produced book *Hydra and the Birth of Experimental Biology—1744* (The Boxwood Press, 1986), were an invaluable source for many of the historical details in this chapter. The Lenhoffs' volume includes engaging accounts of Trembley and his times, his experiments with hydra, and his era's fascination with regeneration. It also contains the Lenhoffs' translation of Trembley's four memoirs, which overflow with Trembley's character as a scientist and his meticulous observations of the small green creatures he was so taken by.

p. 13 **countless little green nubs:** This particular species of freshwater hydra is green because of algae living inside one layer of its cells. Howard Lenhoff, interview by the author, February 9, 2002.

p. 15   **He would cut his "insecte":** *A History of Regeneration Research: Milestones in the Evolution of a Science,* edited by Charles E. Dinsmore (Cambridge/New York: Cambridge University Press, 1991), p. 27. "For example, they used the term 'insect' for any small invertebrate and even for some large vertebrates that appear segmented"—the crocodile, for example.

p. 15   **as Trembley stood witness with his magnifying glass:** In their *Scientific American* article, Vol. 258 (1988), pp. 108-13, Howard and Sylvia Lenhoff made this observation about Trembley: "Amazingly, he made scores of findings in the short span of four years from 1740 to 1744. Moreover, he did it long before the development of sophisticated tools, relying mostly on a magnifying glass and occasionally on a simple single-lens microscope."

p. 16   **". . . one of the heads was immortal and the others almost as bad":** Edith Hamilton, *Mythology* (New York: Mentor Books, The New American Library, 1942), p. 164.

p. 18   **"Apparently these gentlemen have some cherished system":** John R. Baker, *Abraham Trembley of Geneva, Scientist and Philosopher 1710-1784,* (London: Edward Arnold, 1952), p. 43.

p. 19   **Réaumur referred to the animal as a "polyp":** *Hydra and the Birth of Experimental Biology—1744,* p. 5: "Trembley's attitude toward nomenclature appears to have been one of nearly total indifference. He allowed Réaumur, with the collaboration of another colleague, Bernard de Jussieu, to name his animals 'polyps,' with the addition of one ecological ('freshwater') and one morphological ('arms shaped like horns') trait to distinguish them from the octopus and other 'polyps of the sea.'"

p. 19   **"His own eyes and skills":** Charles Dinsmore, "The Foundations of Contemporary Regeneration Research: Historical Perspectives," *Monographs in Developmental Biology,* Vol. 23 (Basel, New York: Karger, 1992), pp. 11-12.

p. 20   **Voltaire was said to be so confident:** *A History of Regeneration Research,* p. 13.

p. 20   **Hoards of naturalists went tromping:** *A History of Regeneration Research,* p. 13.

p. 21   **it may in fact have been Abraham Trembley who first witnessed:** John Baker, Trembley's first biographer, suggests that Trembley was "the first human being to witness cell-division and in the same act the first to see the multiplication of a single-celled plant." *Abraham Trembley of Geneva,* p. 156. Also see pp. 102-3, 154-69. Howard Lenhoff brought this buried piece of history to the author's attention.

p. 22   **"Trembley didn't like theories":** Howard Lenhoff, interview by the author, June 10, 2002.

p. 23   **"In order to extend our knowledge of natural history":** *Hydra and the Birth of Experimental Biology—1744,* Fourth Memoir, p. 187.

2
MOUSE STRAIN 129

Source material includes interviews by the author with Sabine Brauckmann: April 5, 2002. John Gearhart: June 28, 2001; January 8, 2002; August 30, 2003. Barbara Knowles: January 23, 2002. Gail Martin: February 13, 2002. Virginia Papaioannou: February 12, 2002; May 9, 2002. Barry Pierce: February 2, 2002; March 10, 2002. Davor Solter: February 13, 2001. Roy Stevens and Anne Wheeler: January 23, 2002; August 9, 2002. Anne Wheeler: May 7, 2001; July 8, 2002. Donald Varnum: May 7, 2001.

p. 25 **"The stem cell, I believe, is the interpretation of regeneration":** John Hearn, interview by the author, May 18, 2003.

p. 25 **"I felt perfectly free":** Roy Stevens, interview by Susan Mehrtens, The Jackson Laboratory Oral History Collection, June 14, 1986, p. 5.

p. 26 **when he noticed that a chinchilla-colored:** Stevens began his job at the Jackson Laboratory on June 1, 1952, and reports in his oral history (p. 5) that "Maybe it was six months after I had got here" that he saw his first teratoma-burdened mouse. Stevens confirmed to the author (August 9, 2002) that, although the strain has both albino and chinchilla mice, the first one he saw was chinchilla.

p. 26 **Greek for "swollen monster":** "monster" from *teras*, and "swollen" from onk*oma,* from *The Human Teratomas* (New Jersey: Humana Press, 1983), p. 23.

pp. 26-27 **teratomas represent but a small fraction:** Several of Stevens's teratoma papers make for interesting reading. Among them: "Embryology of testicular teratomas in Strain 129 mice," *Journal of the National Cancer Institute*, Vol. 23 (1959), pp. 1249-95; "The biology of teratomas including evidence indicating their origin from primordial germ cells," *Année Biologique* (1962), pp. 586-610; and "The origin and development of testicular, ovarian, and embryo-derived teratomas," *Teratocarcinoma Stem Cells*, Cold Spring Harbor Laboratory, Conferences on Cell Proliferation, Vol. 10 (1983), pp. 23-36.

p. 27 **the first ever described in the testis:** In their paper "Experimental teratoma," *Current Topics in Pathology*, Vol. 59 (1974), pp. 76-77, Ivan Damjanov and Davor Solter note how exceptionally rare spontaneously occurring teratomas are in mammals, including laboratory mice.

p. 27 **indentations shaped like eye sockets:** *The Human Teratomas; Experimental and Clinical Biology*, edited by I. Damjanov, B. Knowles, and D. Solter (New Jersey: Humana Press, 1983), p. 9.

p. 28 **Roy Stevens was instantly captivated by the teratoma:** During the nineteenth century, sightings of these tumors were all the rage in pathology. Writes one historian, "As if competing for the *Guinness Book of Records*, reports of the dermoid with the most teeth [well over 300 . . .], of the greatest weight [over 52 lbs . . .], of the greatest diameter [45 cm], or with different kinds of hair . . . followed." *The Human Teratomas*, p. 2.

p. 29　**an inherited teratoma had never been glimpsed in a mouse:** Stevens's Jackson Laboratory oral history, p. 5. Prior to Stevens's finding, occasional teratomas had been seen in the ovaries of female mice; they might have been inherited, but there was no concrete evidence.

p. 29　**"Prexy" Little was a mouseologist:** Clarence Little acquired his nickname "Prexy" during his years as a college president, for which Prexy was "accepted slang." Martha Harmon, *C.C. Little and the Founding of the Jackson Laboratory*, May 2000.

p. 29　**It was while he was a student at Harvard:** Lee M. Silver, *Mouse Genetics*, adapted for the Web by Mouse Genome Informatics, the Jackson Laboratory. Also, George Snell's online portrait of C.C. Little: *Biographical Memoirs*, Vol. 46 (1975), p. 242. Online version: http://www.nap.edu.

p. 30　**Only one percent of Mouse Strain 129:** The initial paper by Stevens and C.C. Little: "Spontaneous testicular teratomas in an inbred strain of mice," *Proceedings of the National Academy of Sciences*, Vol. 40 (1954), pp. 1080-87. Also, Stevens's Jackson Laboratory oral history, p. 5.

p. 31　**"I could have abandoned the project then and there":** *JAX,* in-house publication of the Jackson Laboratory, Vol. 18 (3), (Fall 1970), p. 6. Many details about Stevens's multiyear attempt to find a mouse teratoma's cell of origin are chronicled in this issue of *JAX* and the Spring 1971 issue, Vol. 19 (1).

p. 31　**"He was decorated several times":** Barry Pierce, interviewed by the author, February 2, 2002. Stevens received the Legion of Merit and the Silver Star; from daughter, Anne Wheeler to author.

p. 35　**"But things turned out":** Driesch's quote appears in Joseph Needham's *Order and Life* (Cambridge, England: Yale University Press, 1936), p. 52. The quote originated in Driesch's "The Science and Philosophy of the Organism," Gifford Lectures (2nd edition, London: Black, 1929). Notes Needham (p. 51), "Considerable astonishment was . . . caused by Driesch's announcement in 1891. . . ."

p. 35　**"prospective potency," as he called it:** *Order and Life*, pp. 53-54. Continues Needham, "This condition of multiple potency of the parts of the early egg-cell has been termed Pluripotence."

p. 36　**"The examination of serial sections":** Francis Herrick, *The American Lobster: A Study of Its Habits and Development* (Washington, DC: Government Printing Office, 1895), p. 108.

p. 38　**"very insignificant complexes of indifferent cells":** Eugen Korschelt, *Regeneration and Transplantation* (Berlin: Gebrudes Borntraeger, 1927); republished by Watson Publishing International, 1990, p. 626.

p. 38　**"Some facts concerning the regressive differentiation":** Henry V. Wilson, "On the behavior of the dissociated cells in hydroids, alcyonaria, and asterias," *Journal of Experimental Zoology*, Vol. 11 (1911), p. 285.

p. 38 **"fine bolting cloth such as is used for tow nets":** Henry Wilson, "On some phenomena of coalescence and regeneration of sponges," *Journal of Experimental Zoology*, Vol. 5 (2), (1907-08), p. 246.

p. 39 **"other animals besides sponges and hydroids":** Henry Wilson, *Journal of Experimental Zoology*, Vol. 11 (1911), pp. 281-87.

p. 39 **"from his single-mindedness":** "Confidential report on candidate for fellowship," the Jackson Laboratory, December 6, 1960.

p. 40 **Thirty percent of his mouse offspring:** Karl Illmensee and Leroy C. Stevens, "Teratomas and chimeras," *Scientific American* (April 1979), p. 123.

p. 40 **"By tracing the testicular teratoma back":** Roy Stevens, *JAX*, Vol. 19 (1), (Spring 1971), p. 13.

p. 41 **"Well we tried, but we couldn't get it to work":** Donald Varnum, interview by the author, May 7, 2001.

p. 42 **"Roy or I formed a rather small, but intensely interested":** Barry Pierce, *Teratomas and Differentiation*, edited by M.I. Sherman and D. Solter (New York: Academic Press, 1975), p. 3.

p. 43 **Kleinsmith and Pierce had:** L.J. Kleinsmith and G.B. Pierce, "Multipotentiality of single embryonal carcinoma cells," *Cancer Research*, Vol. 24 (1964), pp. 1544-51.

p. 43 **Pierce and Dixon uncovered:** G.B. Pierce and F.J. Dixon, "Testicular teratomas, I. Demonstration of teratogenesis by metamorphosis of multipotential cells," *Cancer*, Vol. 12 (1959), pp. 573-83.

p. 43 **"took dogma right by the throat":** Barry Pierce, interview by the author, February 2, 2002.

p. 44 **"The cells were made tumorous simply by activation":** Armin C. Braun, "The reversal of tumor growth," *Scientific American* (November 1965), p. 79. Also of interest by Braun: "Plant cancer," *Scientific American* (June 1952), pp. 66-72.

3

THE PURPLE CELL

Source material includes interviews by the author with Jane Barker: January 23, 2002; January 31, 2002; February 7, 2003. Seldon Bernstein: January 30, 2002; September 6, 2002. William Lensch: November 2003. Ernest McCulloch: July 16, 2001; December 19, 2001; January 9, 2002; February 16, 2003. James Till: June 19, 2001; January 18, 2002; September 5, 2002; December 20, 2002. Irving Weissman: January 11, 2001; February 6, 2002.

p. 47 **"There is a lot of biology between the stem cell and mature cell":** Harvey Patt, paraphrased: "It was Harvey Patt who once said that there is a lot of biology between the stem cell and mature cell." L.F. Lameston, from his "Concluding Address." See *Stem Cells of Renewing Populations*, edited by

A.B. Cairnie, P.K. Lala, and D.G. Osmond (New York: Academic Press, 1976), p. 375. Patt was a radiobiologist at the University of California, San Francisco.

p. 47   **Signs abounded that regenerative cells:** Barry Pierce described (interview by the author, February 2, 2002) the thinking in the 1950s. "Cell biologists knew that basal cells in skin were regenerative tissue, but it wasn't known what happened in many other organs" in regard to tissue renewal. "In the gut, there was nothing that would lead you to believe there were stem cells. Yet the average-sized human being will slough in their feces about forty kilograms of dead gastrointestinal cells every year, and about ten kilograms of dead white blood cells. That means there has to be a tremendous effort to maintain these tissues in their adult functioning state. We're talking about before stem cells were really thought about in cell biology."

p. 48   **had been planted in people's minds back in 1909:** T.M. Fliedner, "Characteristics and potentials of blood stem cells," *Stem Cells,* Vol. 16 (1998), suppl 1, pp. 13-29.

p. 49   **"If they gave 200 rads, that wasn't enough to kill mice":** Irving Weissman, interview by the author, February 6, 2002.

p. 50   **The first effective animal-to-animal bone marrow transplants:** One of the first such papers was E. Lorenz . . . E. Shelton, "Modification of irradiation injury in mice and guinea pigs by bone marrow injections," *Journal of the National Cancer Institute,* Vol. 12 (1951), pp. 197-201.

p. 50   **a British team led by Charles Ford:** C.E. Ford . . . J.F. Loutit, "Cytological identification of radiation-chimaeras," *Nature,* Vol. 177 (1956), pp. 452-54.

p. 51   **a thousand of the laboratory's 100,000 mice had perished:** *JAX,* Vol. 15 (4), (Winter 1967), p. 8.

p. 53   **"we were fumbling at the edge of the known":** Seldon Bernstein, email to the author, April 10, 2004.

p. 53   **Russell's experiment lifted off in 1956:** Seldon Bernstein, interviews by the author, January 30, 2002, and September 6, 2002.

p. 54   **"It was the first rescue for anemia":** Jane Barker, interview by the author, January 23, 2002.

p. 54   **Russell and Bernstein went on to:** S.E. Bernstein and E.S. Russell, "Implantation of normal blood-forming tissue in genetically anemic mice, without X-irradiation of host," *Proceedings of the Society for Experimental Biology and Medicine,* Vol. 101 (1959), pp. 769-73.

p. 54   **"The hypothesis was that stem cells existed":** Ernest McCulloch, interview by the author, July 16, 2001.

p. 57   **"One of the things Ernest and I agreed upon":** James Till, interview by the author, January 18, 2002.

p. 58   **"When I saw those lumps, it occurred to me at once":** Ernest McCulloch, interview by the author, January 9, 2002.

p. 60   **in 1962 Becker produced an elegant set of experiments:** A.J. Becker, E.A. McCulloch, and J.E. Till, "Cytological demonstration of the clonal nature of spleen colonies derived from transplanted mouse marrow cells," *Nature*, Vol. 197 (1963), pp. 452-54.

p. 61   **they could at least propose a definition:** Jim Till believes that the first paper in which his group offered a definition of *stem cell* was: L. Siminovitch, E.A. McCulloch, and J.E. Till, "The distribution of colony-forming cells among spleen colonies," *Journal of Cellular and Comparative Physiology*, Vol. 62 (1963), pp. 327-36. "Lou Siminovitch made crucial conceptual contributions to the work that led to this paper," noted Till in an email to the author.

p. 62   **The current belief is that an average-weighted adult:** *Science News*, June 16, 2001, p. 378.

pp. 62-63   **Till and McCulloch had pocketed:** J.E. Till and E.A. McCulloch, "A direct measurement of the radiation sensitivity of normal mouse bone marrow cells," *Radiation Research*, Vol. 14 (1961), pp. 213-22. The report was reprinted in *The Journal of NIH Research*, Vol. 4 (June 1992), pp. 73-78, with commentary by McCulloch and Till included.

p. 63   **Weissman and his team achieved:** G.J. Spangrude, S. Heimfeld, and I.L. Weissman, "Purification and characterization of mouse hematopoietic stem cells," *Science*, Vol. 241 (1988), pp. 58-62.

p. 64   **The valuable insight that scientists learned by:** See E. Donnall Thomas's Nobel Lecture, "Bone marrow transplantation—past, present and future," December 8, 1990, which is available online at: http://www.nobel.se/medicine/laureates/1990/thomas-lecture.pdf. Thomas cites (p. 579) the late '60s as "the beginning of the 'modern' era of human allogeneic marrow grafting."

p. 64   **In 2002, an estimated 30,000 autologous transplants:** International Bone Marrow Transplant Registry, IBMTR/ABMTR Newsletter, Vol. 10, Issue 1, November 2003. See: http://www.ibmtr.org.

4

MYSTERY IN A DISH

Source material includes interviews by the author with: Carolyn Compton: September 30, 2002. Howard Green: May 1, 2002; September 17, 2002; September 19, 2002; September 25, 2002. James Rheinwald: February 27, 2002; November 1, 2002; November 6, 2002. Pamela Robey: April 23, 2002.

p. 67   **"If you can't grow something, you can't improve its growth, because there's nowhere to start":** Howard Green, interview by the author, May 1, 2002.

p. 67   **"I was doing it for the very reason":** John Gearhart, interview by the author, January 8, 2002.

p. 72   **Historians claim that the first person to successfully graft:** Howard M. Lenhoff and Sylvia G. Lenhoff, "Tissue grafting in animals: Its discovery in 1742 by Abraham Trembley as he experimented with hydra," *Biological Bulletin*, Vol. 166 (February 1984), pp. 1-10.

p. 73   **"I cannot explain what became of the body":** Abraham Trembley, Fourth Memoir, p. 175, Sylvia and Howard Lenhoff, *Hydra and the Birth of Experimental Biology—1744* (The Boxwood Press, 1986).

p. 75   **So successful were they in fine-tuning their methods:** An article by Howard Green in the November 1991 issue of *Scientific American* ("Cultured cells for the treatment of disease," pp. 96-102) provides fuller detail about Green's growing of keratinocytes. Also refer to J.G. Rheinwald and H. Green, "Formation of a keratinizing epithelium in culture by a cloned cell line derived from a teratoma" (pp. 317-30) and "Serial cultivation of strains of human epidermal keratinocytes: The formation of keratinzing colonies from single cells" (pp. 331-44), *Cell*, Vol. 6 (1975).

p. 78   **whose wounds proved more problematic:** G.G. Gallico . . . H. Green, "Permanent coverage of large burn wounds with autologous cultured human epithelium," *The New England Journal of Medicine*, Vol. 311 (1984), pp. 448-51.

5
THE EMBRYO EXPERIMENTS

Source material includes interviews by the author with Marie Di Berardino: January 30, 2004. Alan Bradley: February 4, 2004. Martin Evans: May 9, 2002; February 18, 2003; February 9, 2004. Matthew Kaufman: December 10, 2002. Gail Martin: February 2, 2002; February 13, 2002; March 17, 2004. Virginia Papaioannou: February 12, 2002; May 9, 2002; January 14, 2004. Elizabeth Robertson: April 13, 2004.

p. 83   **"The concept of an embryo":** Scott F. Gilbert, *Developmental Biology* (Sixth edition, Massachusetts: Sinauer Associates, Inc., 2000), p. 3.

p. 83   **The first gene ever identified in an organism:** The LacZ gene of the bacteria *E. coli* was reported in 1969 by Jonathan Beckwith's laboratory at Harvard Medical School.

p. 83   **Nobel laureate François Jacob cogently offered at a scientific meeting:** Jacob, who received a Nobel Prize in 1965 in medicine for his inroads into genes in bacteria, made these comments at a teratoma meeting held at Cold Spring Harbor in 1982: "Concluding remarks," *Teratocarcinoma Stem Cells*, edited by L.M. Silver, G. Martin, and S. Strickland (New York: Cold Spring Harbor Laboratory, 1983), pp. 683-88.

p. 85   **In the late '60s:** Martin Evans, interview by the author, May 9, 2002.

p. 86   **they'd never grown their stem cells long-term:** Pierce, interview by the author, March 26, 2004. Pierce said they grew bits of teratoma, but never attempted to make cell lines with EC cells.

p. 88   **One notable example was organ transplants:** The first heart transplant was done in '67, the patient surviving another eighteen days. That same year the first successful liver transplant was performed by Dr. Thomas Starzl at the University of Colorado Health Sciences Center, Denver. Indicative of how far organ transplantation has come, doctors at the Jackson Memorial Hospital in Miami announced on March 18, 2004, that they had transplanted, as a unit, eight organs into a six-month-old baby with a smooth-muscle disorder. Coralie Carlson, Associated Press, March 18-19, 2004.

p. 89   **Retinoic acid, for instance:** Anton Jetten, director of the Cell Biology Section, National Institute of Environmental Health Sciences, NIH, interview by the author, December 1, 2003.

p. 90   **a longtime question about the embryo had found some answers:** John Gurdon, conversation with the author, October 18, 2002. For further background, see: J.B. Gurdon and J.A. Byrne, "The first half-century of nuclear transplantation," *Proceedings of the National Academy of Sciences* (USA), Vol. 100 (2003), pp. 8048-52. This paper can be found at PNAS Online, Classics of the Scientific Literature: http://www.pnas.org/misc/classics.shtml.

p. 90   **Marie Di Berardino, an eminent developmental geneticist:** John Gurdon, in an email to the author (April 6, 2004), notes the following: "Marie Di Berardino was a member of the Briggs and King research group that first successfully transplanted living cell nuclei to eggs. Marie made a major contribution by discovering the chromosome abnormalities that arise from nuclear transplantation. Over many decades she has contributed greatly to the field of nuclear transfer in amphibia, well before mammalian nuclear transplantation was successful. In particular she has been by far the most important contributor to nuclear transfer work with the American frog Rana pipiens."

p. 91   **Robert Briggs and Thomas King at the Institute for Cancer Research:** Briggs and King began their collaboration in February 1950. Marie Di Berardino, interview by the author, March 29, 2004. See Marie Di Berardino's memoir of Briggs: *Biographical Memoirs,* Vol. 76 (1999), pp. 50-63. Online version: http://stills.nap.edu/html/biomems/rbriggs.html.

p. 92   **would a research team vindicate Gurdon:** K. Hochedlinger and R. Jaenisch, "Monoclonal mice generated by nuclear transfer from mature B and T donor cells," *Nature*, Vol. 415 (2002), pp. 1035-38.

p. 92   **"Briggs and King, and then Gurdon showed us":** Virginia Papaioannou, interview by the author, May 9, 2002.

p. 92 **"No one had ever even guessed":** John R. Baker, *Abraham Trembley of Geneva, Scientist and Philosopher 1710-1784* (London: Edward Arnold, 1952), p. xviii.

p. 95 **Scientists' new found ability to grow:** For more about the keen interest in stem cells from teratomas at the time, and how they might assist biologists in exploring the mammalian embryo and human genetic diseases, see Gail Martin's paper, "Teratocarcinomas and mammalian embryogenesis," *Science*, Vol. 209 (1980), pp. 768-76.

p. 96 **lying in a dish and "marking time":** Martin Evans, interview by Virginia Papaioannou, Albert Lasker Award for Basic Medical Research, 2001: http://www.laskerfoundation.org.

p. 97 **Five months after Evans and Kaufman published:** The Evans-Kaufman study: "Establishment in culture of pluripotential cells from mouse embryos," *Nature*, Vol. 292 (1981), pp. 154-56. Gail Martin's study: "Isolation of a pluripotent cell line from early mouse embryos cultured in medium conditioned by teratocarcinoma stem cells," *Proceedings of the National Academy of Sciences* (USA), Vol. 78 (1981), pp. 7634-38.

p. 97 **Martin Evans persevered toward this goal:** Additional accounts of how Martin Evans went from his early studies of teratoma-burdened mice in the late '60s to engineering a mouse with a genetic modification include: M. Evans, "The cultural mouse," *Nature Medicine*, Vol. 7 (2001), pp. 1081-83. Also of interest: press releases and interviews on the Lasker Foundation's Web site: http://www.laskerfoundation.org.

p. 98 **By '83, they had injected embryonic stem cells:** Allan Bradley, email to the author, April 13, 2004.

p. 98 **Mario Capecchi at the University of Utah:** For a full telling by Capecchi about the milestone of engineering mice that have specific genes knocked out of them, refer to his article "Targeted gene replacement" in *Scientific American* (March 1994), pp. 52-59.

6
THE CANARY'S SONG

Source material includes interviews by the author with Joseph Altman: April 25, 2004. Arturo Alvarez-Buylla: March 13, 2003. Shirley Bayer: April 25, 2004. Fred Gage: April 15, 2002; March 19, 2003. Steven Goldman: November 27, 2001; December 7, 2001; February 17, 2002; May 8, 2002; February 5, 2003. Peter Marler: March 17, 2003; May 14, 2003. Fernando Nottebohm: May 10, 2001; January 9, 2002; February 21, 2003; March 3, 2003; March 4, 2003. Stewart Sell: February 4, 2003.

p. 101 **"Stem cells can divide many times":** Fernando Nottebohm, interview by the author, January 9, 2002.

p. 102   **alongside the stem cell in the bone marrow . . . lived a cousin stem cell:** S.A. Kuznetsov and P.G. Robey, "A look at the history of bone marrow stromal cells: The legacy of Alexander Friedenstein," *Graft*, Vol. 3 (2000), pp. 278-83.

p. 105   **trained as a psychologist, Altman:** Joseph Altman, interview by the author, April 25, 2004.

p. 105   **"The first paper I published was a question mark":** J. Altman, "Are new neurons formed in the brains of adult mammals?" *Science*, Vol. 135 (1962), pp. 1127-28.

p. 107   **She and her Purdue team reported in *Science*:** S.A. Bayer . . . P.S. Puri: "Neurons in the rat dentate gyrus granular layer substantially increase during juvenile and adult life," *Science*, Vol. 216 (1982), pp. 890-92.

p. 107   **Accounts of new neurons forming:** M. Cayre . . . A. Strambi, "The common properties of neurogenesis in the adult brain: From invertebrates to vertebrates," *Comparative Biochemstry and Physiology*, Part B, Vol. 132 (2002), pp. 1-15. R.H. Nordlander and J.S. Edwards, "Postembryonic brain development in the monarch butterfly," *Wilhelm Roux' Archive*, Vol. 164 (1970), pp. 247-60. P.R. Johns and S.S. Easter, "Growth of the adult goldfish eye," *Journal of Comparative Neurology*, Vol. 176 (1977), pp. 331-41.

p. 108   **a student of the field's originator, William Thorpe:** The details about William Thorpe came from Peter Marler, interviewed by the author, March 17, 2003. Marler recounted that the first person to apply the sound spectrograph to the study of birdsong was actually Donald Borror, a distinguished field biologist at Ohio State University. Yet once Thorpe encountered the first of these instruments in England in the late '40s, it was Thorpe, more than Borror, who grasped "its tremendous potential for studying birdsong and took the lead in applying it in a general way," explains Marler. Describes Thorpe in his *Nature* paper titled "The process of song-learning in the chaffinch as studied by means of the sound spectrograph" (March 13, 1954), p. 465, "For the first use of the 'Sonograph' I have been greatly indebted to the kindness and courtesy of the Superintendent and staff of the Admiralty Research Laboratory at Teddington, whose 'Sonograph' was for a long time the sole instrument of its kind in England."

p. 109   **"We all believe it's there":** Fernando Nottebohm, interview by the author, February 21, 2003.

p. 110   **three to four times larger than a female's:** F. Nottebohm, "From bird song to neurogenesis," *Scientific American*, Vol. 260 (1989), p. 75.

p. 110   **From decades past came a handful of reports:** Several of these reports (that testosterone elicits song in female birds) are referenced in Fernando Nottebohm's paper "Testosterone triggers growth of brain vocal control nuclei in adult female canaries," *Brain Research*, Vol. 189 (1980), pp. 429-

36. One paper refers to the pet-store scam: E.H. Herrick and J.O. Harris, "Singing female canaries," *Science*, Vol. 125 (1957), pp. 1299-1300. A different paper by Herrick describes how several months' worth of testosterone injections caused a "broad Breasted Bronze" turkey hen to start "gobbling" like a male. "The influence of androgens in a female turkey," *Poultry Science*, Vol. 30 (1951), p. 758.

p. 111   **It was the first time that a hormone:** Nottebohm's resulting paper: "Testosterone triggers growth of brain vocal control nuclei in adult female canaries," *Brain Research*, Vol. 189 (1980), pp. 429-36.

p. 112   **In various papers, including . . . in *Science* in 1981:** "A brain for all seasons: Cyclical anatomical changes in song control nuclei of the canary brain," *Science*, Vol. 214 (1981), pp. 1368-70.

p. 115   **"I was sure they were neurons because":** Steven Goldman, interview by the author, February 5, 2003.

p. 115   **"There seemed to be no question we were looking at neurons":** Fernando Nottebohm, interview by the author, February 21, 2003.

p. 116   **and in his papers mentioned others before him:** J. Altman, "Autoradiographic and histological studies of postnatal neurogenesis IV," *Journal of Comparative Neurology*, Vol. 137 (1969), p. 433. To quote Altman in that paper: "Several investigators (Allen, '12; Bryans, '59; Globus and Kuhlenbeck, '44; Opalski, '34; Rydberg, '32) reported the presence in adult animals and man of a mitotically active 'subependymal layer' (Kershman, '38) around the ependymal wall of the anterior lateral ventricle."

p. 116   **"These facts would suggest a high rate of cell proliferation":** J. Altman and G.D. Das, "Autoradiographic and histological evidence of postnatal hippocampal neurogenesis in rats," *Journal of Comparative Neurology*, Vol. 124 (1965), p. 329.

p. 117   **They announced their exceptional news:** S. Goldman and F. Nottebohm, "Ependymal neurogenesis in adult female canaries," *Society for Neuroscience Abstracts*, Vol. 8 (140), 1982. F. Nottebohm and S. Goldman, "Connectivity and kinetics of neurons born in adulthood," *Society for Neuroscience Abstracts*, Vol. 8 (140), 1982. Steven Goldman, email to the author, April 19, 2004.

p. 117   **Their paper followed in April:** "Neuronal production, migration, and differentiation in a vocal control nucleus of the adult female canary brain," *Proceedings of the National Academy of Sciences* (USA), Vol. 80 (1983), pp. 2390-94.

p. 121   **They were en route to Jerusalem:** See Nottebohm's "A white canary on Mount Acropolis," which contains the lecture (series: The King Solomon Lectures) that he gave at the Hebrew University, Jerusalem; *Journal of Comparative Physiology*, Vol. 179 (1996), pp. 149-56.

7
AFTER THE BIRTH OF LOUISE BROWN

Source material includes interviews by the author with Ariff Bongso: April 9, 2003; April 24, 2003. Doug Powers: April 3, 2003; April 11, 2003.

p. 123 **"I guess I have never felt that knowledge is scary":** Ralph Brinster, interview by the author, August 20, 2003.

p. 123 **to make better sense of a hierarchy:** Beatrice Mintz, for one, was investigating "a hierarchy of stem cells" in the '70s. See "Gene expression in neoplasia and differentiation," *Harvey Lectures* Series 71 (New York: Academic Press, 1978), p. 217.

p. 124 **Researchers had pried similar cells:** Ariff Bongso's 1994 paper in *Human Reproduction* (Vol. 9, pp. 2110-17) cites rabbits as the first animals from which ES cell lines were obtained, in 1965 (p. 2110). Bongso's paper also includes (p. 2114) several other reports of cell lines made from cells taken from the inner cell mass of other species.

p. 125 **literally, "in the glass" of a petri dish:** Nowadays the term "in vitro," as it applies to in vitro fertilization, is actually a misnomer, given that today's petri dishes are usually made of plastic.

p. 126 **Tens of thousands of test-tube babies later:** According to the American Society for Reproductive Medicine, since 1981 more than a million ART (assisted reproductive technology) babies, which includes IVF babies, have been born worldwide. Of those, over 250,000 were born in the United States.

p. 128 **An estimated two-thirds or more of egg-and-sperm unions:** "What is an embryo?" by H.W. Jones Jr. and L. Veeck, *Fertility and Sterility*, Vol. 77 (April 2002), p. 659, "but a conservative estimate is that at least two-thirds of the products of oocyte and sperm fusion are in some way defective, either chromosomally or perhaps more subtly at the molecular level. The carrier of these abnormalities is so abnormal that it never implants, or, if it does implant, usually perishes very early in development."

p. 128 **A survey published in 2003:** "Cryopreserved embryos in the United States and their availability for research," *Fertility and Sterility*, Vol. 79 (2003), pp. 1063-69.

p. 129 **"that it seemed to come from God's own whisper":** *Asiaweek*, August 24, 2001, p. 34.

p. 131 **"The embryo is not a person":** Lewis Wolpert, email to the author, April 16, 2003. About his well-known maxim, Wolpert communicated (in a second email to the author, November 20, 2002), "I did actually say it to a doctor at a meeting in Belgium who clearly irritated me; and my friend Jonathan Slack—now Prof at University of Bath—wrote it down and published it. I have a poster of it—the actual quote is 'It is not birth, marriage, or death, but gastrulation that is the most important event in your life'. A little exaggeration but even so . . . Lewis."

    p. 131   **Bongso and his crew had carved:** A. Bongso . . . S. Ratnam, "Isolation and culture of inner cell mass cells from human blastocysts," *Human Reproduction*, Vol. 9 (1994), pp. 2110-17.

8
MONKEYS TO HUMANS

Source material includes interviews by the author with Alta Charo: March 28, 2003; March 26, 2004. Norman Fost: December 15, 2002; January 19, 2003; May 2, 2003; March 11, 2004. John Hearn: May 18, 2003; June 7, 2003. Jennifer Kalishman: July 10, 2003; July 22, 2003. Colin Stewart: April 29, 2003. James Thomson: August 10, 2002; May 30, 2003. Michael West: June 2, 2003; July 9, 2003; April 22, 2004.

    p. 133   **"As he told me, he is interested":** From a second-grader's interview with James Thomson that appeared in *Falk Kids Press*, a supplement produced by Falk Elementary School in Madison, Wisconsin, that ran in the *Wisconsin State Journal* on April 14, 2002.

    p. 134   **"In short, the unborn have never been recognized":** *Roe v. Wade*, 410 U.S. 113, January 22, 1973.

    p. 134   **"to systematically address the issue of research on conceptuses":** Lori B. Andrews, "Regulation of experimentation on the unborn," *Journal of Legal Medicine*, Vol. 14 (1993), p. 25.

    p. 135   ***Newsweek* spotlighted the "new hope":** Sharon Begley, "Cures from the womb: Fetal tissue promises new hope for incurable diseases and beguiling questions of science," *Newsweek*, February 22, 1993, pp. 49-51.

    p. 135   **research into transplanting fetal tissue had carried on:** From the NIH's Office of Science Policy: "Chronology—Human Fetal Tissue Research," draft August 29, 2002, footnote 3: "The moratorium also did not affect other types of research (i.e., basic and preclinical research) involving human fetal tissue obtained from induced or spontaneous abortions or stillbirths."

    p. 136   **Each year in the early '90s as many as 1.5 million fetuses:** L.B. Finer and S.K. Henshaw, "Abortion incidence and services in the United States in 2000," *Perspectives on Sexual and Reproductive Health*, Vol. 35 (2003), p. 8.

    p. 140   **As he told an audience years later:** "Stem Cells on Land and at Sea" conference, Mt. Desert Island Biological Laboratory and the Jackson Laboratory, Bar Harbor, August 10, 2002.

p. 140 **to grow a whole plant from a single mature corn cell:** James Thomson, email to the author, June 2003: Mature cell, not a stem cell.

p. 140 **Frederick Steward . . . had been the first to achieve this feat:** Marc Cathey, "Good breeding curbs invasiveness," *The American Gardener*, Vol. 82 (2), (March-April 2003).

p. 141 **as one Michigan zoologist described:** Susan Douglas Hill, "Origin of the regeneration blastema in polychaete annelids," *American Zoologist*, Vol. 10 (1970), pp. 101-12. Hill at the University of Michigan noted in this paper that "the origin of cells of the regeneration blastema has long been a source of controversy," but that "it has become quite well-established that in amphibians the limb blastema"—the knot of cells at the stump of a missing limb—"arises from previously differentiated cells in the immediate vicinity of the amputation site." These cells appeared to undergo "dedifferentiation." Continued Hill, "Gurdon's (1964) revealing experiments with nuclear transplants have shown that in at least some cells the genome is not irreversibly altered or blocked by differentiation but retains its capacity to support development if placed in a suitable environment."

p. 141 **"I was never sure we would survive it":** Davor Solter, "James Thomson and the holy grail," by Leslie Whitaker, *The Pennsylvania Gazette* (Jan-Feb 2002).

p. 142 **"Once you put a blastocyst in culture":** Colin Stewart, interview by the author, April 29, 2003.

p. 142 **"We chatted about how it would be interesting":** James Thomson, interview by the author, August 10, 2002.

p. 143 **Problem solving was right up this Phi Beta Kappa's alley:** Thomson conveyed that he might have ended up a mathematician had it not been for his teacher Frederick Meins at the University of Illinois, who taught Thomson in the honors biology program. Thomson, in notes emailed to the author, July 17, 2003: "Fred certainly has a right to claim me as his intellectual progeny if he cares to, as, if he had not popped up in my life at that point, I probably would have continued to study Mathematics. One of his graduate students went on to make the first transgenic plant, so given the current controversy over GMO organisms in Europe, and the controversy over human ES cells, Fred can really claim to have influenced world events."

p. 146 **"made me think we could outsmart aging":** Michael West, interview by the author, July 9, 2003.

p. 146 **aim of Geron's scientists became . . . telomerase:** Michael West, interview by the author, April 22, 2004. As West describes, Elizabeth Blackburn's lab had found the "footprint" of the telomerase molecule, but it had not been isolated.

p. 148 **Wisconsin Alumni Research Foundation:** Not to be confused with the University of Wisconsin's alumni office, its patent office WARF is a story unto itself. WARF was founded in 1925 as a result of Professor Harry Steenbock's then-controversial idea of providing the university with the patent rights to his discovery of how to activate vitamin D in food products. His innovation virtually eliminated the bone disease rickets while raking in millions of dollars in patent profits. WARF went on to fund the research of university scientists from proceeds derived from a growing pool of lucrative patents. See "Commercializing university research," *The New England Journal of Medicine*, Vol. 314 (1986), pp. 1621-26.

p. 148 **"a week later Mike West visited":** James Thomson, email to the author, July 17, 2003.

p. 152 **how can its beginning be any less important:** New scientific evidence in 2002 would support this pro-life stance, in Doerflinger's opinion, which was the finding that within twenty-four hours of conception fertilized mammalian eggs have body-plan features previously undetected, and so weren't just blobs of cells. See *Nature*, Vol. 418 (July 4, 2002), p. 14.

p. 152 **"In his great novel *The Brothers Karamazov*":** Joseph A. Fiorenza, letter to Congress, July 10, 2001. U.S. Catholic Bishops, Office of Communications: http://www.nccbuscc.org/prolife/issues/bioethic/stemcell 71001.htm.

9
EPIC UPON EPIC

Source material includes interviews with Marc Cathey: August 12, 2003. Peter Donovan: May 21, 2004. David Gardner: July, 1, 2003. John Gearhart: April 9, 2001; June 11-13, 2001; June 28, 2001; January 8, 2002; November 6, 2002; May 23, 2003; August 3, 2003; August 30, 2003. Robert Griesbach: December 11, 2003. Jeffrey Jones: September 8, 2003. Bruce Lahn: April 18, 2003. Beatrice Mintz: November 2, 2003; November 11, 2003. Virginia Papaioannou: May 9, 2002. Michael Shamblott: April 23, 2001; June 13, 2001; September 3, 2003. Susan Singer: April 10, 2002. Michael West: April 22, 2004.

p. 155 **"The field of developmental biology has changed dramatically":** Virginia Papaioannou, interview by the author, December 12, 2002.

p. 155 **Professor Shunong Li:** Information about Shunong Li was provided by Bruce Lahn, a geneticist at the University of Chicago and an acquaintance of Li. Bruce Lahn, interview by the author, April 18, 2003; and emails to author, May 22, 2003, and August 7, 2003. According to Lahn, Li started to derive human embryonic cells in 1997.

p. 155 **Mediterranean anemia, an inherited blood disease:** Alternative names include thalessemia and Cooley's anemia. This group of inherited conditions interfere with the production of hemoglobin, leading to too few red blood cells.

p. 156   **Christopher Graham at Oxford:** In an email (June 6, 2003) to the author, Graham stated that the title of his first stem cell-related "licence application [1987] was 'Derivation of cell lines from the human conceptus to investigate the growth regulation of embryonic and tumour cells for the development of effective preimplantation diagnosis.'" Graham said in an earlier email (June 5, 2003) to the author, "It is possible that there seemed to be more people trying to derive human ES cells than were actually doing so. I obtained my material from two IVF clinics and consequently had to go through two ethical cttees in two different hospitals."

p. 156   **"their inner cell masses . . . screamed stem cells":** Robert Edwards, email to the author, May 20, 2003.

p. 157   **a roughly six-day-old embryo:** Jeffrey Jones, email to the author, February 13, 2004: "Actually, the very 1st stem cell line was created from a DAY 7 embryo. The other 4 cell lines came from DAY 6 embryos."

p. 158   **The very term *embryology* had been growing stale:** Developmental biologist J.D. Ebert, in a talk at the Taniguchi Symposium on Developmental Biology, April '97, remarked that "the term 'developmental biology' did not come into common usage until the early 1950s." From the symposium catalogue, p. 86.

p. 158   **larger-than-life Cecil B. De Mille figure:** Marc Cathey, who worked in Frederick Steward's lab, provided this depiction of Steward in a telephone interview by the author, August 12, 2003. He recalled Steward as saying "we only do epics" and further described Steward as "all drama-and-hell."

p. 158   **Steward's precedent of cloning a carrot plant:** According to Cathey, Steward's lab-grown carrot was the first plant cloned from a single cell. See *The American Gardener* (March/April 2003), p. 13. Author could find no further confirmation. Over time, botanists have found that Steward's technique of growing a plant from a single adult cell is less efficient than growing a plant from a group of meristem cells taken from an adult plant's organ, according to Robert Griesbach of the U.S. National Arboretum. The latter is the method most often used nowadays for cloning plants. One of Steward's claims to fame, however, is his use of the special coconut-milk culture. The vast majority of blooms that greet one at a flower market have been grown in a similar coconut-milk medium, confirms Griesbach, which means that "Steward's experiment is part of every birth, wedding, and funeral," says Cathey, interview by the author, August 12, 2003. Robert Griesbach, interview by the author, December 11, 2003; and email, December 15, 2003.

p. 159   **"possesses a power of its own":** Theodor Schwann, *Microscopical Researches into the Accordance in the Structure and Growth of Animals and Plants,*" translated from German by Henry Smith (London: The Sydenham Society, 1847), p. 192. (Of interest: pp. 186-215.)

pp. 159-160   **Robert Briggs and Thomas King had set the stage:** R. Briggs and T.J. King, "Transplantation of living nuclei from blastula cells into enucleated frogs' eggs," *Proceedings of the National Academy of Sciences* (USA), Vol. 38 (1952), pp. 455-63.

p. 160   **Beatrice Mintz . . . and Tibby Russell . . . had determined:** "Gene-induced embryological modifications of primordial germ cells in the mouse," *Journal of Experimental Zoology*, Vol. 134 (1957), pp. 227-37.

p. 161   **"invisible developmental history":** Beatrice Mintz, interview by the author, September 11, 2003.

p. 167   **"the collaboration was really a hedge":** Michael West, interview by the author, June 2, 2003.

p. 171   **Ian Wilmut and his team at the Roslin Institute:** I. Wilmut . . . K.H. Campbell, "Viable offspring derived from fetal and adult mammalian cells," *Nature*, Vol. 385 (1997), pp. 810-13.

p. 172   **They included "Huxleyish scenarios":** *Johns Hopkins Magazine*, November 1997.

p. 173   **"If we are successful in genetically manipulating our cells":** John Gearhart, notes emailed to the author, July 30, 2003.

p. 173   **"When Dolly happened":** James Thomson, interview by the author, May 30, 2003.

p. 174   **Thomson and Jeffrey Jones finally started growing:** Jeffrey Jones, in an email (September 22, 2003) to the author: "My records also show that the first human embryo donated to Jamie's IRB protocol was donated on 5/10/96 but not thawed and cultured until a few months later on 8/23/96. For the next year and a half (from this date until December '97), an additional 32 embryos were donated to this protocol. However no hES [human ES] cell lines were created until we switched to David Gardner's sequential embryo culture system in January of 1998."

p. 174   **Working in a small room:** James Thomson's early IVF embryo work, done in a small room at the University of Wisconsin Hospital, was isolated from federally-funded projects. The research later moved off-campus to WiCell Research Institute. As Alta Charo cites, a ruling in the fall of '01 would verify that, in fact, it is not illegal to do private work in the same lab as federally-funded work. Rather, the problem that presents itself, says Charo, is that it's so logistically difficult to comply with the legal need to keep every petri dish and pipette purchased with private funds separate from materials purchased with public funds "that many researchers choose to create a separate location."

p. 174   **the two papers were published:** J.A. Thomson . . . J.M. Jones, "Embryonic stem cell lines derived from human blastocysts," *Science*, Vol. 282 (1998), pp. 1145-47. M.J. Shamblott . . . J.D. Gearhart: "Derivation of pluripotent stem cells from cultured human primordial germ cells," *Proceedings of the National Academy of Sciences* (USA), Vol. 95 (1998), pp. 13726-31.

p. 175   **Professor Li lost his cell line:** Bruce Lahn, in an email to the author, May 22, 2003: "Professor Shunong Li published his human ES cell line in January of 1998, in *Academic Journal of Sun Yat-Sen University of Medical Science* 19(1):77-78. He is still at the University, and I have frequent contact with him. In the paper, he passaged his line 5 times, and showed that his cells are positive for alkaline phosphotase, a marker for undifferentiated human ES cells. The line was lost when, during summer vacation, the liquid nitrogen tank dried up."

p. 176   **Gearhart observed in a 1999 letter to Roy Stevens:** Letter from John Gearhart to Roy Stevens, March 16, 1999.

p. 176   **"There is almost no realm of medicine":** Harold Varmus quoted by Sharon Schmickle, *Star Tribune*, April 11, 1999.

p. 177   **"Research and demonstrations of clinical efficacy":** Paul Berg, testimony, U.S. Senate Committee on the Judiciary, "Drawing the line between ethical regenerative medicine research and immoral human reproductive cloning," March 19, 2003.

10
MARROW TO BRAIN?

Source material includes interviews by the author with Alejandro Sánchez Alvarado: March 15, 2002; January 27, 2004. Diana Clarke: April 15, 2002; April 18, 2002. George Daley: May 29, 2001; June 21, 2001; August 16, 2002; October 3, 2003; January 19, 2004. Curt Freed: October 24, 2003. Mark Keating: May 14, 2003. Jordana Lenon: July 30, 2003. Éva Mezey: April 24, 2002; September 23, 2003. Timothy Moffitt: May 31, 2001. Shannon Odelberg: March 10, 2004. Charles Peters: October 13, 2003. Catherine Verfaillie: May 20, 2002; November 4, 2003. Frances Verter: November 13, 2003.

p. 179   **"But extraordinary claims require extraordinary evidence":** Carl Sagan, *Broca's Brain: Reflections on the Romance of Science* (New York: Random House, 1979), p. 62.

p. 180   **they "could transform the practice of medicine":** George Daley, "Stem cells: Debunking the myths, developing the medicine," Museum of Science, Boston, March 14, 2001.

p. 180   **during a press seminar at Whitehead:** "New Horizons in the Post-Genome World," a Whitehead Institute Press Seminar, May 2-4, 2001. Daley's talk "Stem Cells: Medicine and Myth" was on May 4, 2001.

p. 182   **Harvard Medical School's Evan Snyder:** B.D. Yandava . . . E.Y. Snyder, "Global cell replacement is feasible via neural stem cell transplantation: Evidence from the dysmyelinated *shiverer* mouse brain," *Proceedings of the National Academy of Sciences* (USA), Vol. 96 (1999), pp. 7029-34.

p. 182   **scientists at the Salk Institute in California and Sweden's Sahlgrenska University Hospital:** P.S. Eriksson . . . F.H. Gage, "Neurogenesis in the adult human hippocampus," *Nature Medicine*, Vol. 4 (1998), pp. 1313-17.

p. 182   **London taxi driver study:** E.A. Maguire . . . D.C. Frith, "Navigation-related structural change in the hippocampi of taxi drivers," *Proceedings of the National Academy of Sciences* (USA), Vol. 97 (2000), pp. 4398-403.

p. 183   **Weissman's Stanford lab had persevered:** G.J. Spangrude . . . I.L. Weissman, "Purification and characterization of mouse hematopoietic stem cells," *Science*, Vol. 241 (1988), pp. 58-62. C.M. Baum . . . B. Peault, "Isolation of a candidate human hematopoietic stem-cell population," *Proceedings of the National Academy of Sciences* (USA), Vol. 89 (1992), pp. 2804-08.

p. 184   **"We *can* do this":** John Gearhart, Oncology Seminar Series, Dana-Farber Cancer Institute, April 10, 2001.

p. 185   **the first adults to benefit from this new technology:** Two adults with chronic myelogenous leukemia. See: A.L. Pecora et al., "Prompt and durable engraftment in two older patients with high risk chronic myelogenous leukemia (CML) using ex vivo expanded and unmanipulated umbilical cord blood," *Bone Marrow Transplantation*, Vol. 25 (2000), pp. 797-99. Frances Verter, writer and keeper of the outstanding Web site "A Parent's Guide to Cord Blood Banks," noted the following in an email (November 13, 2003) to the author: "The small volume of cord blood collections, and hence the insufficient number of stem cells, was once a barrier to adult usage, but this has now been overcome by either multiplying the cells in vitro prior to transplant (J Jaroscak, et al., *Blood* 2003; 101:5061), or by transplanting multiple cord blood units simultaneously (JR Wingard, Medscape [TM]2001; Medscape Conf Covereage, Amer Soc Hem 43rd Annual Meeting)." See http://www.parentsguidecordblood.com for more information on the subject. Also refer to issue #51 of the newsletter on the Blood & Marrow Transplant Information Network (http://www.bmtinfonet.org), another top-notch Web site.

p. 186   **how effectively can a baby's banked umbilical cord blood treat an illness:** For in-depth information about cord blood's medical uses and how to evaluate the public/private banks that collect it, visit "A Parent's Guide to Cord Blood Banks" at http://www.parentsguidecordblood.com. This Web site provides insights into the pros and cons of cord blood banking, the diseases treated with cord blood, and how the stem cells in cord blood compare with those in bone marrow.

p. 188   **"We really had a hard time convincing ourselves":** Angelo Vescovi, quoted by John Travis, "Cellular conversion turns brain into blood," by John Travis, *Science News*, January 23, 1999. The reported study: C.R.

Bjornson . . . A.L. Vescovi, "Turning brain into blood: A hematopoietic fate adopted by adult neural stem cells in vivo," *Science*, Vol. 283 (1999), pp. 534-37.

p. 189   **in the lab of Éva Mezey:** "Turning blood into brain: Cells bearing neuronal antigens generated in vivo from bone marrow." Ran in the December 1, 2000, issue of *Science*, Vol. 290.

p. 190   **claim by Helen Blau and coworkers at Stanford:** "From marrow to brain: Expression of neuronal phenotypes in adult mice." Ran in the December 1, 2000, issue of *Science*, Vol. 290.

p. 190   **A collaboration orchestrated by Diane Krause:** D.S. Krause . . . S.J. Sharkis, "Multi-organ, multi-lineage engraftment by a single bone marrow-derived stem cell," *Cell*, Vol. 105 (2001), pp. 369-77.

p. 190   **This finding was soon eclipsed by another:** Catherine Verfaillie first announced the finding of her bone marrow stem cells at a meeting of the American Society of Hematology, December 1999. Her lab would publish several papers about the nature of these cells. One of the most notable: "Pluripotency of mesenchymal stem cells derived from adult marrow," *Nature*, Vol. 418 (2002), pp. 41-49.

p. 191   **A journal article had lately reminded scientists:** The article was by Darwin Prockop, then at Allegheny University of the Health Sciences in Philadelphia: "Marrow stromal cells as stem cells for nonhematopoietic tissues," *Science*, Vol. 276 (April 4, 1997), pp. 71-74.

p. 193   **This seemed to strongly validate her cell's multipotent versatility:** Verfaillie called her cell a "multipotent adult progenitor cell," or MAPC.

p. 193   **Numerous studies now indicate:** Two such studies showing the brain's recruitment of new cells soon after injury: H. Nakatomi . . . M. Nakafuku, "Regeneration of hippocampal pyramidal neurons after ischemic brain injury by recruitment of endogenous neural progenitors," *Cell*, Vol. 23 (2002), pp. 429-41. J. Liu . . . F.R. Sharp, "Increased neurogenesis in the dentate gyrus after transient global ischemia in gerbils," *Journal of Neuroscience*, Vol. 18 (1998), pp. 7768-78.

p. 196   **"Are there in this Insect":** *History of Regeneration Research*, p. 74. Original source: V.P. Dawson, "Nature's enigma: The problem of the polyp in the letters of Bonnet, Trembley, and Reaumur," American Philosophical Society, Philadelphia, 1987.

p. 196   **an excerpt from one lengthy report:** T.H. Morgan, "The physiology of regeneration," *Journal of Experimental Zoölogy*, Vol. 3 (1906), pp. 458-59. Before long, Morgan would become so discouraged by his failure to figure out regeneration, that he would switch to studying genetics and wind up just about the most famous geneticist of his day.

p. 197   **Either these immature cells were holdovers:** In his two-volume work *Regeneration and Transplantation* (1927, 1931), Eugen Korschelt frequently cites these two possible explanations for limb replacement. In one passage, he refers to the "redifferentiation of cells or reactivation of available reserves" (p. 2). Elsewhere he cites these prerequisites for the regrowth of lost parts: "Sufficient indifferent cellular material must be available for this, be it that cells which have actually remained indifferent are present, being left over from embryonic development, and which proceed to divide when the occasion arises, or that it is possible for already differentiated cells to return to the indifferent state" (p. 620). Some scientists of his day, Korschelt notes, felt that "once cells have developed along specific lines, they are in no case able later to change tack and to continue along different lines" (p. 334). Korschelt, however, maintains that "in view of the facts at our disposal, there can scarcely be any denying the possibility that developmental processes are reversible" (p. 335).

p. 197   **"Either it's a very early stem cell":** Catherine Verfaillie, interview by the author, May 20, 2002.

p. 198   **Max Planck Institute team in Dresden:** shows cells in salamander's tail dedifferentiating into embryonic cells: K. Echeverri and E.M. Tanaka, "Ectoderm to mesoderm lineage switching during axolotl tail regeneration," *Science*, Vol. 298 (2002), pp. 1993-96.

p. 198   **"The ability to regrow missing parts":** Alejandro Sánchez Alvarado, interview by the author, April 15, 2002.

p. 199   **"Any cell in the body can behave as any other cell":** Christopher Potten, conversation with the author, March 24, 2003.

p. 200   **Mark Keating . . . and his postdoc Shannon Odelberg:** suggest that specialized mammalian cells can dedifferentiate: S.J. Odelberg . . . M.T. Keating, "Dedifferentiation of mammalian myotubes induced by msx1," *Cell*, Vol. 103 (2000), pp. 1099-1109.

p. 200   **"I don't think there's any reason":** Mark Keating, interview by the author, May 14, 2003.

p. 201   **George Daley and Rudolf Jaenisch collaborated:** W.M. Rideout . . . R. Jaenisch, "Correction of a genetic defect by nuclear transplantation and combined cell and gene therapy," *Cell*, Vol. 109 (2002), pp. 17-27.

p. 201   **On August 9, 2001, George W. Bush . . . to announce:** For press release, see http://www.whitehouse.gov/news/releases/2001/08/print/20010809-2.html.

p. 201   **Thomson himself "was reportedly hang gliding".** Jordana Lenon, interview by the author, July 30, 2003.

p. 202   **The August decision was neither an executive order:** Lana Skirboll, email to the author, July 20, 2004.

p. 203   **He had received an experimental transplant:** Curt Freed, interviews by the author, October 24, 2003; March 2004.

11
THE ART OF CELL REPLACEMENT

Source material includes interviews by the author with Douglas Cotanche: March 19, 2002. Marie Filbin: May 3, 2004. Rebecca Folkerth: October 9, 2001. Donald Gash: January 15, 2003. Emanuela Gussoni: January 7, 2003. Kenneth Kenyon: November 1, 2003; January 16, 2004. Ronald McKay: November 11, 2001; January 31, 2002; January 20, 2004. Donald Orlic: February 20, 2004. Thomas Reh: January 4, 2002; January 15, 2002; January 13, 2004. Simon Tran: October 22, 2003. Ann Traynor: February 2, 2004. Henriette van Praag: January 5, 2004. Michael Young: March 29, 2001; April 3, 2001; March 20, 2002; October 9, 2003; January 26, 2004.

p. 205 **"The challenge is to move basic cell biology":** George Daley, lecture: "Stem cells: Debunking the myths, developing the medicine," Boston's Museum of Science, March 14, 2001.

p. 207 **The University of Toronto researchers:** V. Tropepe . . . D. van der Kooy, "Retinal stem cells in the adult mammalian eye," *Science*, Vol. 287 (2000), pp. 2032-36.

p. 209 **In his talk—which was titled "Is the potential of stem cells":** Ruben Adler, lecture, Schepens Eye Research Institute, May 17, 2001.

p. 211 **Parkinson's patient who had traveled to China:** The journal paper that reviews this case: R.D. Folkerth and R. Durso, "Survival and proliferation of nonneural tissues, with obstruction of cerebral ventricles, in a Parkinsonian patient treated with fetal allografts," *Neurology*, Vol. 46 (1996), pp. 1219-25.

p. 213 **notes Kenneth Kenyon, who developed limbal cell transplants:** K.R. Kenyon and S.C. Tseng, "Limbal autograft transplantation for ocular surface disorders," *Ophthalmology*, Vol. 96 (1989), pp. 709-22.

p. 216 **FDA approval cannot be taken as a true indicator:** Paul Richards, interview by the author, March 8, 2004; email to the author, April 20, 2004.

p. 221 **Approximately one in every 400 to 500 children:** National Institute of Diabetes and Digestive and Kidney Disease (NIDDK), NIH. See http://diabetes.niddk.nih.gov/dm/pubs/statistics/index.htm#8.

p. 221 **"Eighty percent of patients":** James Shapiro, email to the author, January 28, 2004. Shapiro also noted in his email that "over 300 patients have received islet transplants worldwide since the introduction of the Edmonton Protocol. An international multicenter trial is currently underway at the nine sites, funded by the Immune Tolerance Network."

p. 225 **Henriette van Praag at the Salk Institute made a discovery:** H. van Praag . . . F.H. Gage, "Running increases cell proliferation and neurogenesis in the adult mouse dentate gyrus," *Nature Neuroscience*, Vol. 2 (1999), pp. 266-70. Also see: H. van Praag . . . F.H. Gage, "Running enhances neurogenesis, learning and long-term potentiation in mice," *Proceedings of National Academy of Sciences* (USA), Vol. 96 (1999), pp. 13427-31.

p. 226   **It refers to mouse studies conducted by Donald Orlic:** One such Orlic study is: "Bone marrow cells regenerate infarcted myocardium," *Nature*, Vol. 410 (2001), pp. 701-05.

p. 226   **In recent investigations:** Strauer et al., *Circulation*, Vol. 106 (2002), pp. 1913-18; Assmus et al., *Circulation*, Vol. 106 (2002), pp. 3009-17; Stamm et al., *Lancet*, Vol. 361 (2003), pp. 45-46; Tse et al., *Lancet*, Vol. 361 (2003), pp. 47-49.

p. 229   **"Being able to control transitions":** Ron McKay, interview by the author, April 23, 2002.

p. 229   **Two years later, they went a step further:** J.H. Kim . . . R. McKay, "Dopamine neurons derived from embryonic stem cells function in an animal model of Parkinson's disease," *Nature*, Vol. 418 (2002), pp. 50-56.

p. 229   **"up to a fifty percent reduction":** M. Barinaga, "Fetal neuron grafts pave the way for stem cell therapies," *Science*, Vol. 287 (2000), p. 1422.

p. 230   **They derived these regenerative progenitor cells:** Charles Vacanti, emails to the author, March 22 and March 23, 2004. Charles Vacanti is one of four brothers who together are known for pioneering the field of tissue engineering. They have been called the "stem cells for the entire field."

p. 231   **stem cells from sheep bone marrow:** T.E. Perry . . . J.E. Mayer, "Bone marrow as a cell source for tissue engineering heart valves," *The Annals of Thoracic Surgery*, Vol. 75 (2003), pp. 761-67.

p. 231   **approximately 2.5 million people living:** International Campaign for Cures of Spinal Cord Injury Paralysis.

p. 233   **researchers have identified a patch of cells:** These experiments were done by Michael Zuber in the laboratory of William Harris at the University of Cambridge.

12
AFTER TOMORROW

Source material includes interviews by the author with Ariff Bongso: April 9, 2003. Timothy McCaffrey: December 19, 2003. Andrew McMahon: March 29, 2004. Douglas Melton: March 5, 2002; November 20, 2002. Terry Mock (Champion Tree Project): April 4, 2004. Jonathan Tilly: March 29, 2004.

p. 235   **"We are now in the center of biology itself":** Ronald McKay, "Medical evolution" by Gregg Easterbrook, *The New Republic*, March 1, 1999, p. 20.

p. 235   **a study by Japanese scientists that involved planarians:** F. Cebria . . . K. Agata, "FGFR-related gene nou-darake restricts brain tissues to the hea region of planarians," *Nature*, Vol. 419 (2002), pp. 620-24.

p. 235   **shown by Thomas Hunt Morgan:** "Experimental studies of the regeneration of Planaria maculata," *Wilhelm Roux' Archiv für Entwicklungsmechanik der Organismen*, Vol. 7 (1898), pp. 364-97.

p. 236   **the Champion Tree Project International:** For more information visit: http://www.championtreeproject.org.

p. 236   **a Massachusetts General Hospital team unfurled its surprise finding:** F. Johnson . . . J.L. Tilly, "Germline stem cells and follicular renewal in the postnatal mammalian ovary," *Nature*, Vol. 428 (2004), pp. 145-50. "It's true that new markers very much helped with this work, though we were led into this by studying cell death." Jonathan Tilly, interview by the author, March 29, 2004.

p. 236   **what has tended to be "over-hyped":** Christopher Potten, conversation with the author, March 24, 2003.

p. 237   **Possibly as many as ten percent of a cell's genes:** Ron McKay, interview by the author, April 23, 2002.

p. 237   **stridently opposed to human reproductive cloning:** Many researchers who oppose the reproductive cloning of humans nevertheless support the cloning of nonhuman animals. Cloning cows and sheep, an alternative to the lengthy process of breeding, is seen as a valuable source of agricultural products; and, if made to carry foreign genes, cloned animals might be useful producers of important human proteins. Cloning might also help to increase the numbers of certain endangered species. Note that the world's first such clone, a male banteng, was born at the San Diego Zoo in January 2004.

p. 238   **In the United States, scientists . . . who rely on government funding:** The National Institutes of Health alone "supports an extramural research community of over 200,000 research personnel who are affiliated with approximately 1,700 organizations." Lana Skirboll, director of the Office of Science Policy, NIH, email to the author, April 1, 2004.

p. 240   **Singapore's bioethics committee:** "Ethical, legal and social issues in human stem cell research, reproductive and therapeutic cloning," June 2002, http://www.bioethics-singapore.org/resources/reports1.html.

p. 241   **a South Korean team reported:** W.S. Hwang . . . S.Y. Moon, "Evidence of a pluripotent human embryonic stem cell line derived from a cloned blastocyst," *Science*, Vol. 303 (2004), pp. 1669-74.

p. 242   **They maintain that they have developed culture techniques:** Aside from TriStem, other groups that have reported turning one type of mature cell into another includes a team at the Scripps Research Institute in San Diego that announced in 2004 that it has a chemical that can induce muscle cells from mice to dedifferentiate into immature status, from which state the cells can be directed into other cell types.

p. 242   **The British company TriStem:** I.S. Abuljadayel, "Induction of stem cell-like plasticity in mononuclear cells derived from unmobilised adult human peripheral blood," *Current Medical Research and Opinion*, Vol. 19 (2003), pp. 355-75.

p. 243   **"If we say any cell has the potential":** Harold Varmus, "Medical evolution," by Gregg Easterbrook, *The New Republic*, March 1, 1999, p. 23.

p. 243   **"The Dolly experiment says that any cell":** James Thomson, interview by the author, May 10, 2003.

p. 243   **"It would be surprising if the first lines were any good":** Doug Melton, interview by the author, March 5, 2002.

p. 244   **Massachusetts had passed a law in 1974:** Massachusetts General Laws, Chapter 112, Section 12J.

p. 245   **grown in the company of mouse cells:** Andrew McMahon, interview by the author, March 29, 2004.

p. 247   **Author's note:** When I asked Hans Bode whether a hydra really is an immortal animal, his response was: "Well, it might be, but obviously you can't prove it's immortal."

Glossary

Autologous vs. allogeneic transplant   In an autologous transplant, a patient receives back his or her own cells, whereas in an allogeneic transplant a patient receives someone else's cells. The advantage of the first procedure is that a patient doesn't run the risk of an immune system rejection.

Blastocyst   The early mammalian embryo before it implants in the uterus. In human development, the blastocyst forms on day 4-5 after fertilization; in mouse development, on day 3-4. The blastocyst's round sphere consists of an outer layer of cells and an inner cavity that contains embryonic stem cells. The inner stem cells mostly disperse or differentiate within a few days. The blastocyst stage, therefore, is a key time for harvesting stem cells.

Cell line   A continuously dividing batch of cells that originated from one individual. Consider a cell line made from a mammal's embryonic stem cells. One cell or dozens are removed from the blastocyst and placed in the right culture conditions, which ensures that the cells will keep dividing indefinitely. Thus the cells are "immortalized." When a cell line isn't being used, it is frozen; when it's needed, it's thawed and grown again. The advantage of a cell line is that it provides researchers with easy access to a source of unlimited numbers of cells from one individual that are all genetically similar.

Chimera   An animal composed of cells from two or more geneti-cally different animals. Technically speaking, even a person who has received a bone marrow transplant from someone other than an iden-tical twin can be considered a chimera. In biology, the chimeras of-ten utilized are chimeric mice, which are made when cells from one mouse embryo are added to cells of another mouse embryo. The two sets of genes can be distinguished with markers, providing insights into how cells mature and specialize.

Cloning   Creating a genetic duplicate of an organism. In the mod-ern sense, cloning an animal consists of transferring the nucleus of a single cell into an egg cell. This reprograms the nucleus. Its genes lose their previous identity and return to a potent state that results in the growth of a developing embryo.

> **Therapeutic**   If a clone is grown only a few days, solely for the sake of getting it to blastocyst stage to obtain its stem cells, this amounts to *therapeutic cloning*, which is also called *research cloning*.

> **Reproductive**   But if a clone is transferred into the uterus and grown to birth, this is considered *reproductive cloning*. Al-though animals have been reproduced this way, no human clone has been confirmed born.

Differentiate   When a cell advances from a less mature to a more mature state. Cells that differentiate, via their progeny, are headed toward specialization. In some cases a cell can *de*differentiate, or re-vert to a less specialized state.

Embryonal carcinoma cell—EC cells   Stem cells derived from a teratoma type of tumor. Once in culture, they can be called EC cells.

Embryonic germ cells—EG cells   Precursors to egg and sperm cells found in the developing fetus. Once in culture, they can be referred to as EG cells.

Embryonic stem cells—ES cells   Immature, unspecialized cells in the inner cell mass of an early embryo. Once in culture, they can go by the nickname ES cells.

Genome   An organism's full set of genes; its DNA. In the case of humans, every cell contains the entire genome except for red blood cells. A cell in your kidney, for instance, contains the human genome's approximately 30,000 genes, as does a skin cell or a heart cell.

Germ cells   Egg and sperm cells. These germinal cells in the reproductive organs populate the *germ line*, which is simply the lineage consisting of egg/sperm cells that perpetuate a species.

Hematopoietic stem cell   An adult stem cell that gives rise to the specialized red and white blood cells of the blood and immune systems. These cells are found in a child's umbilical cord blood as well as an adult's bone marrow.

Implantation   When the embryo implants in the mother's uterus. In mice this happens around day 4 after fertilization; in humans, around day 7-8.

In vitro fertilization (IVF)   Fertilizing an egg outside of the body with sperm in order to create an embryo. The fertilized embryo may be transferred into a woman or frozen and stored.

In vitro vs. In vivo   Cells grown *in vitro* are cells that, having been removed from an individual, are grown in a specially prepared culture media in a dish. Cells *in vivo* are cells as they exist in a living system. When tissue sections are looked at through a microscope, their cells, although dead, are seen in the context of a natural *in vivo* environment.

Knockout mouse A mouse that has one or more genes inactivated. This is done by altering a gene in an embryonic stem cell and then putting the altered stem cell into an early mouse embryo, whereupon the progeny cells will populate many of the developing tissues. Knockout mice serve as models that mimic many human genetic diseases.

Lineage The pathway that cells follow, by way of countless divisions and stages of differentiation, as they advance from an immature state to a specialized state. The further down this pathway they go, the more they "commit" to a specific "fate," or tissue.

Mesenchymal stem cell (Also called a marrow stromal cell) An adult stem cell found in the bone marrow that gives rise to bone, fat, cartilage, and muscle cells.

Multipotent An adult stem cell that produces more than one specialized cell type, as opposed to a *unipotent* stem cell that gives rise to just one type of specialized cell. For instance, a hematopoietic stem cell is multipotent; a sperm stem cell is unipotent.

Neural stem cell An adult stem cell that resides in central nervous system tissue and generates that tissue's cell types: neurons and glia. Glial cells consist of astrocytes and oligodendrocytes.

Neurogenesis The birth of new neurons in the central nervous system.

Parthenote An egg cell that begins cleavage and development like an embryo, but without fertilization by sperm. A product of *parthenogenesis*.

Plasticity A cell's ability to change its differentiation state, or even its cell type. Cells have *plasticity,* for instance, if they can switch between pluripotent and mature, or mature and pluripotent, or even from one lineage to another, as from blood stem cell to muscle stem cell.

Pluripotent   A cell that can generate every cell type found in the embryo and adult, but not the cells of structures that support the embryo.

Progenitor   A "child" or "grandchild" of a stem cell. A progenitor cell is more differentiated than a stem cell and yet has some stemness that can permit it to generate multiple types of mature cells. Unlike a stem cell, it lacks the unlimited ability to self-renew.

Stem cell   An unspecialized cell that is capable of generating, via its progeny, differentiated cells and/or more cells like itself (stem cells). A stem cell ostensibly got its name because of its unique ability to give rise to different cell types, sometimes throughout the life of an organism. In contrast, when a specialized cell divides, it generates more of itself for a limited number of divisions.
  1. Embryonic stem cell   Stem cells in the early embryo. This early stem cell appears to have the greatest potential to generate a range of differentiated cells.
  2. Adult stem cell   Stem cells in the differentiated tissue of developing and full-grown organisms. Several different organs contain "adult" stem cells, including the bone marrow, liver, intestine, brain, and skin. Because an adult stem cell is slightly more differentiated than an embryonic stem cell, it appears to have less potential. Like all stem cells, it can make more of itself as well as generate specialized cells, but only those of the tissue it inhabits.

Stemness   A stem cell's potential to generate multiple mature cell types.

Totipotent   A cell that has the ability to differentiate into all cell types. In mammals, only the zygote and the cells resulting from its first few divisions have this all-encompassing ability, which includes being able to generate cells that make up the umbilical cord and placenta.

Transdifferentiate   When a cell from one tissue switches lineages and becomes a cell of another tissue. An example would be a bone marrow cell changing to a skin cell.

Zygote   The cell that results from a sperm cell fertilizing an egg cell. The zygote—the fertilized egg—is the beginning of development.

Acknowledgments

Some years ago, Jane Nevins of the Dana Press pointed me in the direction of writing a book on stem cells. Then there was no turning back—all thanks to Jane's contagious interest in the subject. I am deeply indebted, as well, to my editor Jeffrey Robbins at Joseph Henry Press who acquired this book, as he did my last one. It has meant a great deal to be believed in and to have the chance to write on subjects that are changing medicine. My agent Doe Coover, meanwhile, is a Rock of Gibraltar of good advice. She has a real knack for putting things in perspective.

The research for this book relied heavily on a great many scientists. Were it not for their firsthand accounts of their experiments and surrounding events, I would have been hard-pressed to gather and understand many essential aspects of their work. In particular, I owe a tremendous amount to Diana Clarke, George Daley, John Gearhart, Steve Goldman, Willy Lensch, Gail Martin, Ron McKay, and Virginia Papaioannou for patiently fielding my unending questions and generously giving of their knowledge and time. I am grateful as well for my conversations with four special gentlemen from Canada: Howard Green, Ernest McCulloch, Barry Pierce, and James Till. Each contributed largely to these pages, each heard from me a lot, and each was a gentleman to the end. Several members of the Jackson Laboratory family were extremely helpful as well: Anne Wheeler and Roy Stevens, Barbara Knowles, Jane Barker, Joyce Peterson, Muriel Davisson, and librarian Douglas Macbeth.

My appreciation extends to numerous others, each of whom provided details and explanations over the course of many conversations and emails: Joseph Altman, Shirley Bayer, Marie Di Berardino, Seldon Bernstein, Hans Bode, Alta Charo, Charles Dinsmore, Martin Evans, Norman Fost, Fred Gage, David Gardner, Scott Gilbert, John Gurdon, Jeffrey Jones, Jennifer Kalishman, Matthew Kaufman, Mark Keating, Kenneth Kenyon, Diane Krause, Doug Melton, Éva Mezey, Beatrice Mintz, Douglas Powers, Thomas Reh, James Rheinwald, Pamela Robey, Stewart Sell, Michael Shamblott, Susan Singer, Evan Snyder, James Thomson, Catherine Verfaillie, Fran Verter, Irving Weissman, Michael West, and Leonard Zon. In addition, without the invaluable assistance of Fernando Nottbohm and Michael Young, the chapters in which they appear would have been thin and inadequate. Gratitude as well goes to John Hearn in Australia and Ariff Bongso in Singapore, both of whom were wonderfully helpful and added further dimension to the book.

Although Abraham Trembley was beyond reach, some of the most enjoyable and enlightening moments of my research were spent on the phone or in the company with his successors Howard Lenhoff, Sylvia Lenhoff, Hans Bode, and Richard Campbell. Thanks to Campbell at UC/Irvine, I even got to look through a microscope and know what it's like to create two individuals from one with "a snip" of the scissors. (Trembley pulled off far harder dissections with a lesser lens.)

Stephen Mautner, Jeff Robbins, Ann Merchant, Dick Morris, Robin Pinnel, Rachel Marcus, and others at the National Academy of Sciences' Joseph Henry Press work unceasingly to bring the inspiring world of scientific discovery alive for readers, and its authors are tremendously fortunate to have this roof over our heads. This book also benefitted from Toni Gardner's astute line editing; Paula Henry's perceptive comments; Nancy Kurtz's eye and Pentax; Lenny Golay's enthusiasm; Rudy Tanzi's emails and general support; and last but not least, Pierrette Coadou's newspaper clippings and friendship.

Cheers to other special friends: Jessie Bethoney, Chris Brown, Madeline Drexler, Julie and Bayard Henry, Susan Hunnewell, Mary Myers Kauppila, Ted Kurtz, Sonia and David Landes, Jeff Munger, Bill Patrick, Bonnie Shershow, Betsey Vinton, Rob Whitman, and also my paddle tennis chums and Brooklin, Maine, family and cohorts. Love and appreciation to my mother Katharine Rodgers, Christopher Rodgers, Arrel Linderman, and Bob Linderman for bearing with my book work. You can call me in the mornings now.

Index

protocol, 77
tissue replacement for large-area
burns, 77-82
Slack, Jonathan, 262
Smith, Christopher, 249
Smith, Lois, 53
Smithies, Oliver, 98
Snails, 20
Snyder, Evan, 9, 182
Society for Neuroscience, 117
Solter, Davor, 41, 141, 142
Songbirds, vocal learning in, 107-108
Sound spectrograph, 108, 260
Spallanzani, Lazzaro, 19, 20, 205
Specter, Arlen, 238
Spinal cord injuries, 11, 231-232
Sponges, 38-39, 198-199
Staats, Joan, 30
Stanford University, 49, 63, 183, 190,
245
Starfish, 26
Starzl, Thomas, 258
Steenbock, Harry, 265
Stem cells. *See also* Adult stem cells;
Embryonic stem cells;
Hematopoietic stem cells
cloning and, 90-93
in cultures, 89, 266
dedifferentiation, 139, 159, 171,
179-180, 188, 264, 278
definition of, 61, 72, 256, 281
differentiation, 86, 93, 94, 114, 278
in hydra, 73-74
malignant, 42-43, 179-180, 199,
246; *see also* Embryonic
carcinoma cells; Teratomas
medical potential of, 8-12, 88-89,
93-94, 120-121, 123, 147, 156,
176-177, 180, 181, 184, 187-
188
multipotency, 5, 60, 280
organ-dwelling, 4
as origin of disease, 245-246
origin of term, 35
in plants, *see* Meristem cells

pluripotency, 3-4, 32, 35, 42-43, 67,
85, 89, 96, 123, 132, 253, 281;
see also Embryonal carcinoma
cells; Embryonic stem cells
self-renewing, 2-3
totipotency, 3, 35, 39, 92, 159, 189,
282
transdifferentiation of, 188-189, 282
types, 2
Stentors, 21
Steptoe, Patrick, 125, 156
Stevens, Jean, 32-33
Stevens, Leroy ("Roy"), 25-26, 28-34,
35, 39-42, 43, 44, 46, 47, 51, 67,
68, 84, 85-86, 87, 141, 147, 157,
162-163, 175, 176, 194, 252,
253
Steward, Frederick, 140, 158-159, 185,
266
Stewart, Colin, 172
Stingray, 107
Stratified squamous epithelium, 70
Sun Yat-Sen University, 155
Szentágothai, John, 189

T

T Breeders, 6-7
Tanzi, Rudolph, 213
Tarkowski, Andrzej, 93, 160
Teeth, 224
Telomerase, 146, 153, 264
Telomeres, 146, 147, 165, 168
Teratomas, 47, 82, 99, 252, 253
characteristics, 26-28
chemotherapy, 45-46
culturing, 67, 85-86
differentiation process, 68, 86, 88,
157
differentiation therapy, 43, 44-45,
139
EC cells, 27, 33, 43, 44, 51, 67, 88,
139, 147
ES cells and, 144